V-31g

Inv.-Nr. 33053

Geographisches Institut
der Universität Kiel
ausgesonderte Dublette

Geographisches Institut
der Universität Kiel
Neue Universität

Inv.-Nr. 33053

Sea and Coast

NATIONAL
ATLAS
  OF
SWEDEN

Sea and Coast

SPECIAL EDITOR

Björn Sjöberg

THEME MANAGER

Swedish Meteorological and
Hydrological Institute

The National Atlas of S

SNA Publishing will publish between 1990 and 1996 a government-financed National Atlas of Sweden. The first national atlas, *Atlas över Sverige* was published in 1953–71 by *Svenska Sällskapet för Antropologi och Geografi, SSAG* (the Swedish Society for Anthropology and Geography). The new national atlas describes Sweden in seventeen volumes, each of which deals with a separate theme. The organizations responsible for this new national atlas are *Lantmäteriverket, LMV* (the National Land Survey of Sweden), *SSAG* and *Statistiska centralbyrån, SCB* (Statistics Sweden). The whole project is under the supervision of a board consisting of the chairman, Jim Widmark and Sture Norberg (LMV), Staffan Helmfrid and Åke Sundborg (SSAG), Frithiof Billström and Gösta Guteland (SCB) and Leif Wastenson (SNA). To assist the board and the editors there is a scientific advisory group of three permanent members: Professor Staffan Helmfrid (Chairman), Professor Erik Bylund and Professor Anders Rapp. A theme manager is responsible for compiling the manuscript for each individual volume. The National Atlas of Sweden is to be published in book form both in Swedish and in English, and in a computer-based version for use in personal computers.

The English edition of the National Atlas of Sweden is published under the auspices of the *Royal Swedish Academy of Sciences* by the National Committee of Geography with financial support from *Knut och Alice Wallenbergs Stiftelse* and *Marcus och Amalia Wallenbergs Stiftelse*.

The whole work comprises the following volumes (in order of publication):

MAPS AND MAPPING
THE FORESTS
THE POPULATION
THE ENVIRONMENT
AGRICULTURE
THE INFRASTRUCTURE
SEA AND COAST
CULTURAL LIFE, RECREATION AND TOURISM
SWEDEN IN THE WORLD
WORK AND LEISURE
CULTURAL HERITAGE AND PRESERVATION
GEOLOGY
LANDSCAPE AND SETTLEMENTS
CLIMATE, LAKES AND RIVERS
MANUFACTURING, SERVICE AND TRADE
GEOGRAPHY OF PLANTS AND ANIMALS
THE GEOGRAPHY OF SWEDEN

CHIEF EDITOR	Leif Wastenson
EDITORS	Staffan Helmfrid, Scientific Editor
	Märta Syrén, Editor of *Sea and Coast*
	Ulla Arnberg, Editor
	Margareta Elg, Editor
PRODUCTION	LM Maps, Kiruna
ECIAL EDITOR	Björn Sjöberg
RANSLATOR	Nigel Rollison
GRAPHIC DESIGN	Håkan Lindström
LAYOUT	Typoform/Peter Johnsson, Gunnel Eriksson
EPRODUCTION	LM Repro, Luleå
MPOSITION	Bokstaven AB, Göteborg
TRIBUTION	Almqvist & Wiksell International, Stockholm
R ILLUSTRATION	Lennart Broborn/N

tion

Italy 1992

0-04-5 (All volumes)
16-9 (Sea and Coast)

Contents

"The Lone-Flier Screams..." 6
ARNE GADD

The Oceans 8
JOHAN RODHE

Sea to the West and to the East 10
STIG FONSELIUS

Development of the Seas 14
INGEMAR CATO AND BERNT KJELLIN

Pre-Quaternary 14
Quaternary 14

Bottoms and Sediments 16
INGEMAR CATO, BERNT KJELLIN AND KJELL NORDBERG

Erosion and Deposition 22
What the Shelf Sediment Reveals 22
Microfossils 24

Coast and Shore 26
JOHN O. NORRMAN

Coastal Regions 28
The Archipelago Coast of the Skagerrak 29
The Lowland Coast of the Kattegat 30
The Open Coasts of Skåne 31
The Blekinge Archipelago 32
Kalmarsund, Öland and Gotland 32
The Archipelago Coast of Östergötland 34
The Archipelago Coast of Stockholm and Södermanland 35
The Coastal Plain of the Bothnian Sea 36
The High Coast—Höga Kusten 37
The Coast of the Northern Quark 38
The Archipelago Coast of the Bothnian Bay 39

Weather and Climate of the Sea 40
WEINE JOSEFSSON AND HALDO VEDIN

The Sea and Sea-water 46
JOHAN RODHE

Distibution of Different Water Masses 46
Ice in the Sea 48
Currents and Circulation 48
Variations in Water Level 51

Elements in the Water—Mixing and Dispersion 52
ELISABET FOGELQVIST, STIG FONSELIUS AND LARS-ARNE RAHM

Water in the West and Water in the East 56
LARS ANDERSSON, BERTIL HÅKANSSON, JAN-ERIK LUNDQVIST, ANDERS OMSTEDT, LARS-ARNE RAHM, BJÖRN SJÖBERG AND JONNY SVENSSON

Water Turnover, Fresh Water Supply and Salinity 56
Temperature and Ice 58
Wind Waves 62
Water Level 64
Currents 66
Oxygen and Hydrogen Sulphide 68
Nutrients 70

Life in the Sea 73
HANS CEDERWALL, LARS EDLER, LARS HERNROTH, ALF JOSEFSON, BENGT SJÖSTRAND, JARL-OVE STRÖMBERG, BJÖRN TUNBERG AND INGER WALLENTINUS

Phytoplankton 78
Zooplankton 81
Vegetation 84
Bottom Fauna 88
Fish 94

Marine Resources 96
KJELL GRIP

Boundaries of the Sea 98
STIG CARLBERG AND BJÖRN SJÖBERG

Fisheries and Aquaculture 100
BENGT SJÖSTRAND

Fisheries—an Industry with Ancient Traditions 100
Swedish Fisheries During the 20th Century 101
Fish Processing 108
Leisure-time Fishing 110
Aquaculture 112

Shipping and Navigation 114
KJELL GRIP

Environmental Effects 115

Energy 116
KJELL GRIP

Emissions 118
KJELL GRIP

Mineral Resources 120
INGEMAR CATO AND BERNT KJELLIN

The Changing Sea 122
ANDERS STIGEBRANDT

"The Lone-Flier Screams..."

The sea was never far from my Grandfather's home in Blekinge, glistening between the pine trees or suddenly shining like a blue tablecloth between green bushes, slapping against the small boats at their moorings, whispering in the rushes which closed tightly around us at the innermost end of our cove when we went to the village shop. At sunset a cargo schooner might enter the cove and make for the quayside to load stone from the neighbouring quarries, its sails silhouetted against the reddening sky and hanging loosely in the dying breeze, and their reflection dancing in oily arabesques with the listless motion of the vessel.

This was the sea in its role as an idyllic summer paradise for us mortals, with the humming of insects and the salty tang of the sea mixed with wafts of milk and droppings from cattle grazing the water-meadows. This was the sea showing its friendliest profile; shallow sandy beaches where the sun burnt into one's back, where the water was tepid and soft on the skin, and small dabs tickled the soles of our feet as we cautiously waded along the shore.

But the sea also had another face, darker and more frightening but which I had yet to experience. Old, widely-travelled seamen in the cottages nearby could tell fearful tales of sea monsters, wrecks and ghosts – the tales they could tell! I would be afraid of ghosts for weeks, or even months, afterwards. These old sea-dogs, these global travellers formerly in the service of the Crown, who had seen both negroes, Hottentots and cannibals, made a lasting impression on me as to the vastness of the sea.

I myself was able to confirm that the sea was indeed large. I only needed to climb up on the rock at the end of Grandfather's garden to see it spread out in front of me like a never-ending plain of water. Schooners, ketches and sloops were busily plying to and fro out there – their well-filled sails bluish in colour against the light, yet suddenly blackening beneath the shadow of a cloud hurrying over the silver-glittering expanse, soon to be erased by a grey squall of rain, and yet soon to reappear as if in the developing tray of a photographer. And farthest out on the horizon, where the smoke from steamers lay like forgotten mourning crepe, floated Hanö's low, rather flattened peak, as if it had been thrown into eternity. Usually, we saw it just as a light blue shadow against the light, almost shiny, strip of sky just above the horizon, yet sometimes in intense black. During warm summer evenings after a hot day it disappeared completely in the sea haze – later to reassert itself in the deepening dusk when the revolving eye of the lighthouse started winking.

It was, of course, natural that such sights would make an impression on my playing as a child. I had a whole fleet of Atlantic steamers and clipper-ships (roughly hewn pieces of wood with baking-powder tins as funnels or pieces of wrapping-paper stuck on twigs as sails) with which I explored the oceans in a water-filled old quarry nearby. I also had my own lighthouse, a discarded stable lantern with a stump of candle which I placed on the rock which was my normal vantage point and which I lit when the Hanö lighthouse started to flash its message.

But the influence of these summer experiences in the Blekinge archipelago were to extend into my life as an adult. The island out there on the deserted expanse of the sea would never be obliterated from my thoughts. Every time I looked out over the sea it was there. Everything else on the surface of the water was moving over the expanse of water before disappearing: the shadows of the clouds, the clear-blue ripples on the surface in the calm of the morning, as well as the breakers during gales – not to mention the ships on their journeys to far-off places and countries. But the island was always there at the same place, always just as appealing and just as impossible to reach. It was not surprising that in later years I suffered from an acute obsession with islands. And as I expressed it in a book, "the absolute necessity to hunt the continuously disappearing horizons". The seed that was sown during those summers was to germinate and grow in the neighbourhood of sea and ships since we moved house almost down to the harbour itself in Göteborg. But it was certainly here, in that Baltic cove, that the seed was first planted.

Probably whole libraries have been written on the yearning for the sea and the extraordinary attraction of the sea's horizon, that the English call sea fever. Philosophical studies have been made of its psychological causes and its effect on the individual. But, first and foremost, it has had an enormous importance for the expansion and development of human civilization. The seas were the unifying link and it was to a large extent over the seas that languages and cultures dispersed. The Mediterranean is one example, the North Sea is another, the Atlantic a third. And behind each new expedition by sea were not just economic reasons, but also this remarkable force: the longing to discover what is hidden beyond the horizon.

If we now keep to the personal level, we can summarize in one single sentence what I would like to call – for the sake of simplicity – the sea syndrome. An Anglo-Saxon poem dating back to the seventh or eighth century makes mention of it, but nothing during later centuries that has been written about these things has been able to express with such power what the anonymous poet, or should we say bard, then achieved. He describes his voyages over the seas and all the toils and hardships meted out by the cruel sea. He had experienced

Koster fjord in northern Bohuslän

both cold and storm and loneliness on the northern seas. Nonetheless, as long as he was ashore, on dry land, he was entirely helpless and unable to resist the temptations of what, in reality, are frightening and hostile elements: "the lone-flier screams, resistlessly urges the heart to the whale-way over the stretch of seas".

This brings to mind an old man who might have recognized these words if he had known about them, but expressed the same feelings yet in another way. He was from Öckerö in the southernmost archipelago of Bohuslän and had been a fisherman throughout his life. It was hardly a "lone-flier" that led him to choose his profession – not consciously in any event – but more of an environmentally-conditioned tradition, the influence from his surroundings and the necessity to earn a living. But nonetheless, the allurement of the sea and the open spaces must have been within him, as a dream or in any event as something which cannot be expressed, and which accompanied him throughout his life.

It was fairly long ago that he gave up working on the fishing trawlers but he still had his small boat in which he went out and fished for his everyday requirements. But as he grew older it became increasingly difficult for him to use his boat. His relatives were worried about him and suggested that it was time to sell the boat and remain ashore – who knows, one day he might not even make it home. The old man had a prompt answer: "If you haven't got a boat, then you haven't got anything".

My impression of Bohuslän since the age of about seven was that the coastal fishing population were orientated towards the sea so completely that one might believe that the rest of the country behind the actual shoreline meant little or nothing to them. There seemed to be an invisible boundary between the people who lived with the sound of the sea and waves ringing in their ears and the agricultural population of the hinterland, well rooted in the rich soils of the valleys. The age-old differences in conditions for the ways of life of the two categories certainly played a role, in the same way as the historically conditioned conflicts between land-owners and land-less, between freeholder farmers and cottagers along the shore.

Naturally, as a child I did not understand such conditions but my feeling for shades of meaning gave results. When I moved from the inland to the coast, and vice versa, it was like moving from one kind of people to another. The difference could be observed in many ways, not least in attitudes to the surrounding world.

As late as in those years, that is to say during the 1930's, there were still elderly farmers in the area in which we lived during the summers, who had spent their entire lives within the parish boundaries, possibly with the exception of the conscription training period at Backamo or even a trip to Göteborg. In general, the cottagers along the coasts felt no such restriction in space. The "fields" of the professional fisherman in Bohuslän included not only the waters in the immediate vicinity along the coast but also those in the Skagerrak, which then widened into the North Sea which, in his opinion, might have had a status similar to that held by a farmer for fields lying farthest away. Their "outfields" were, in the same way, the far-off seas around Orkney, Shetland and the Faeroes, and which were finally swallowed up in the open spaces of the Atlantic.

The Oceans

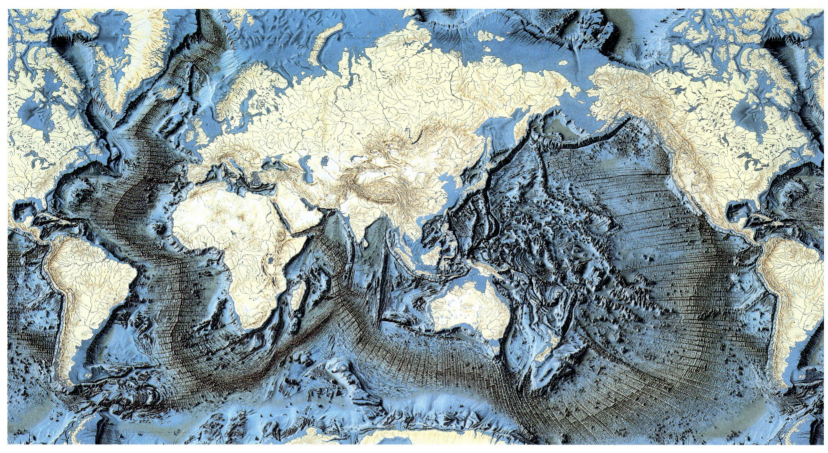

Mountain formations on ocean floors are even greater than those on land.

The sea covers 71% of the surface of the world but with an uneven distribution, the southern hemisphere being 81% covered by water and the northern 61%. The seas of the world are divided into three oceans; the Pacific Ocean, the Atlantic Ocean and the Indian Ocean. These oceans are connected around the Antarctic. The waters around the Antarctic are called the Antarctic Ocean. The northern Atlantic and the northern Pacific are linked through the Bering Strait. The Pacific Ocean is also linked to the Indian Ocean through the islands of Indonesia. The various oceans also have more or less adjacent seas; examples being the Mediterranean, North Sea and Northern Polar Sea belonging to the Atlantic.

Topography

The radius of the world is 6,370 km and all height differences are contained within a 20 km thick layer. This implies that if we compare the Earth with an apple, then the highest mountain on land and the deepest trench in the sea will find space within its thin peel. The surface of the Earth can be divided into two classes according to the elevation, the continents and the ocean floor. Almost half the surface of the Earth is hidden in the sea at depths between 3 and 6 km.

The oceans are crisscrossed by an enormous mountain formation, the mid-ocean ridge. The mid-ocean is divided in the middle by a fissure zone and crossed by faults. The abyssal plains extend on both sides of the ridge. These plains are interspersed with smaller mountainous formations and, in addition, numerous mountain groups are dispersed over their surfaces. The ocean formations reach up to the surface of the sea at only a few places. One example is Iceland, which is part of the mid-ocean ridge in the Atlantic.

The distribution of land and water on the globe's surface is uneven.

CONTINENTAL HEMISPHERE   MARITIME HEMISPHERE

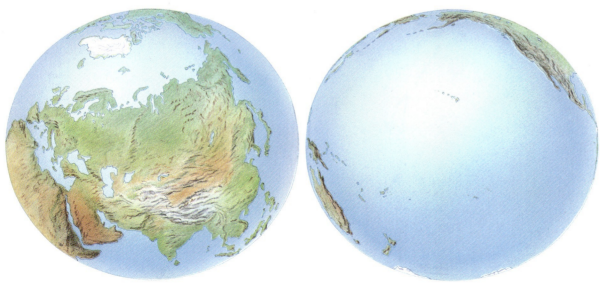

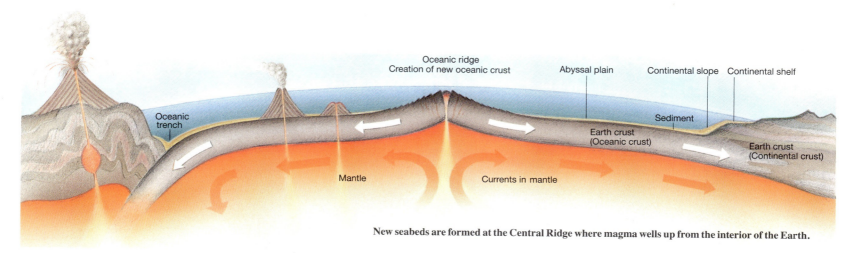

New seabeds are formed at the Central Ridge where magma wells up from the interior of the Earth.

Beyond the abyssal plains, the bottom of the sea rises steeply towards the continents. In many places, the land is surrounded by shallow marine areas with depths down to 200 m. These areas are called continental shelves and are part of the continents. The North Sea-Skagerrak is an example. The deepest parts, the deep sea trenches, are found close to the borders of the oceans.

The Earth's crust consists of a number of plates which move in relation to each other. The mutual rates of movement are of a magnitude of 1–10 cm/year. These movements depend on convection currents in the mantle which are driven by heat development in the centre of the Earth. In places where the plates are moved apart by upwelling magma, a new crust is formed. The mid-ocean ridge in the oceans develops in this way.

The oceanic trenches form in areas where plates collide. If this occurs at the edge of a continent, then a mountainous chain is also formed on land. If the collision instead takes place out in the sea, not only trenches are formed but also chains of volcanic islands. Most trenches are found in the western Pacific, where the deepest of them, the Marianer trench, is also found, with a depth of more than 11 km.

The bottom of the sea is usually covered by loose sediment. This can be divided into terrigenous or pelagic sediments. The terrigenous sediments are found mostly in the proximity of the continents and largely consist of weathering and erosion products which enter the sea from the land via rivers and the atmosphere. Volcanic activity also contributes to the terrigenous sediments. The pelagic sediments are formed in the sea and are of biological (organic) or inorganic origin. Organic sediment consists of shells of phyto- and zooplankton. In areas with high biological production, the organic sediments largely consist of siliceous residues of shells (from *radiolaria* in equatorial areas, and from *diatoms* at higher latitudes). Lime-rich sediments (from *foraminifera* and *pteropoda*) dominate in areas with lower surface production. However, the lime becomes dissolved at low temperatures and high pressure and thus calciferous sediment is not found in the deeper parts of the oceans, where instead we find the inorganic oceanic clay.

The thickness of the sedimental layer is generally at a maximum closest to the continents and decreases out towards the central parts of the oceans. This reflects the formation pattern of the ocean's bottom, the oldest bottom being farthest from the formation zone, the mid-ocean ridge. Far out in the deepest parts of the oceans, the sediment accumulates at a rate of 1–10 mm/1,000 years. On the continental shelf, where terrigenous sediment dominates, the rate of accumulation may be 1 mm/year or higher. In some areas, the currents or the bottom condition may be such that no sediment can accumulate.

BOTTOM SEDIMENT IN THE ATLANTIC OCEAN
- Continental shelf
- Ice rafted
- Carbonate
- Siliceous
- Red clay
- Terrigenous
- Siliceous/Red clay

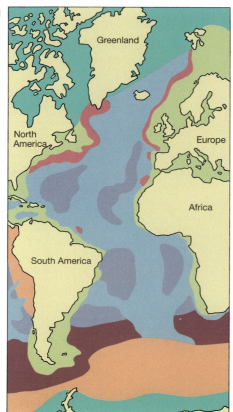

Sediment distribution is mainly determined by the distance to the coast, sea depth and production conditions at the sea surface. (G1)

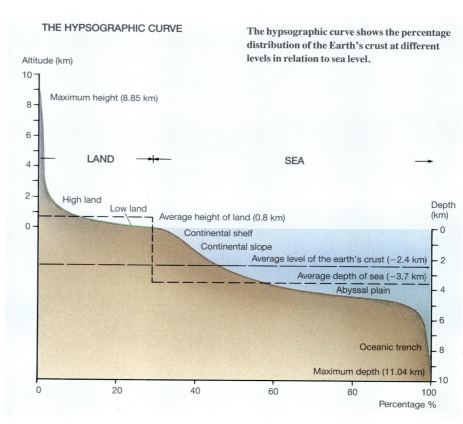

The hypsographic curve shows the percentage distribution of the Earth's crust at different levels in relation to sea level.

Sea to the West and to the East

The Skagerrak, with good links to the North Sea, is the deepest of the marine areas around Sweden. The boundary between the Skagerrak and the North Sea runs between Hanstholm in Denmark and Lindesnäs in Norway. The deepest part of the Skagerrak is the Norwegian Trench (Norska rännan) which extends in towards the Swedish coast along the southern coast of Norway. The greatest depth, the Skagerrak Deep (Skagerrakdjupet), is about 700 m. The Norwegian Trench turns to the south off the Swedish coast and enters the Kattegat. Here it is called the Deep Furrow (Djupa rännan). At the border with the Skagerrak, which is usually drawn along a line from Skagen to the Paternoster Islands, off Marstrand, the depth is about 100 m but it decreases to 70 m off Vinga. There is no sill between the Kattegat and the Skagerrak to prevent exchange of deep water.

The depth in the Kattegat decreases to the south, but there are isolated deeps with depths of up to 100 m. The Sound and the Belt Sea are usually given the common name of the Danish Sounds (Danska sunden). The actual sill to the Baltic in the Sound follows a line from Malmö over the Island of Saltholm and on to Denmark. On the Swedish side, in the Flint Channel (Flintrännan), the sill depth is about 7 m, and on the Danish side, in the Drogden Channel (Drogdenrännan), it is 8 m. The true sill to the Baltic Sea is, however, in the southern part of the Belt Sea between Darsser Ort and Gedser. Here the sill depth is 18 m.

The Baltic Sea, or Eystrasalt as it was known in Ancient Scandinavian, is the world's largest brackish water sea area, with a total surface, including the Danish sounds, of 372,000 km². It can be divided into a number of different areas; the Baltic Proper, the Gulf of Bothnia, the Gulf of Finland and the Gulf of Riga. The two latter do not have hydrographically important sills, and therefore may be considered as bays of the Baltic Proper. The southern part of the Baltic Proper consists of a number of different basins, the Arkona Basin (Arkonabäckenet), the Bornholm Basin (Bornholmbäckenet), Stolpe or Slupsk Channel (Stolpe- or Slupskrännan) and Gdansk Basin, which are all separated from each other by shallows. To the north, is the large Central Basin of the Baltic. With its 249 m, the Gotland Deep (Gotlandsdjupet) is the deepest part East of Gotland. The greatest depth in the Baltic Sea, the Landsort Deep (Landsortsdjupet), 459 m, is off Landsort in the North-west past of the Central Basin and is a crescent shaped fault with very steep walls. The greatest depth West of Gotland is the Norrköping Deep (Norrköpingsdjupet), 205 m.

The Gulf of Bothnia is divided into five areas, the Åland Sea, the Archipelago Sea, the (Northern) Quark, the Bothnian Sea and the Bothnian Bay. The Åland Sea consists of two basins, the Southern Åland Sea and

Ocean limits determined by all coastal states around the Baltic Sea, the Kattegat and the Skagerrak. (G2)

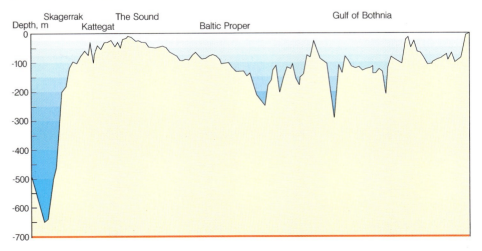

A section from the Skagerrak to the Gulf of Bothnia. The depth scale is enlarged about 1,500 times in comparison with the length scale. The true length of the section is about 2,300 km.

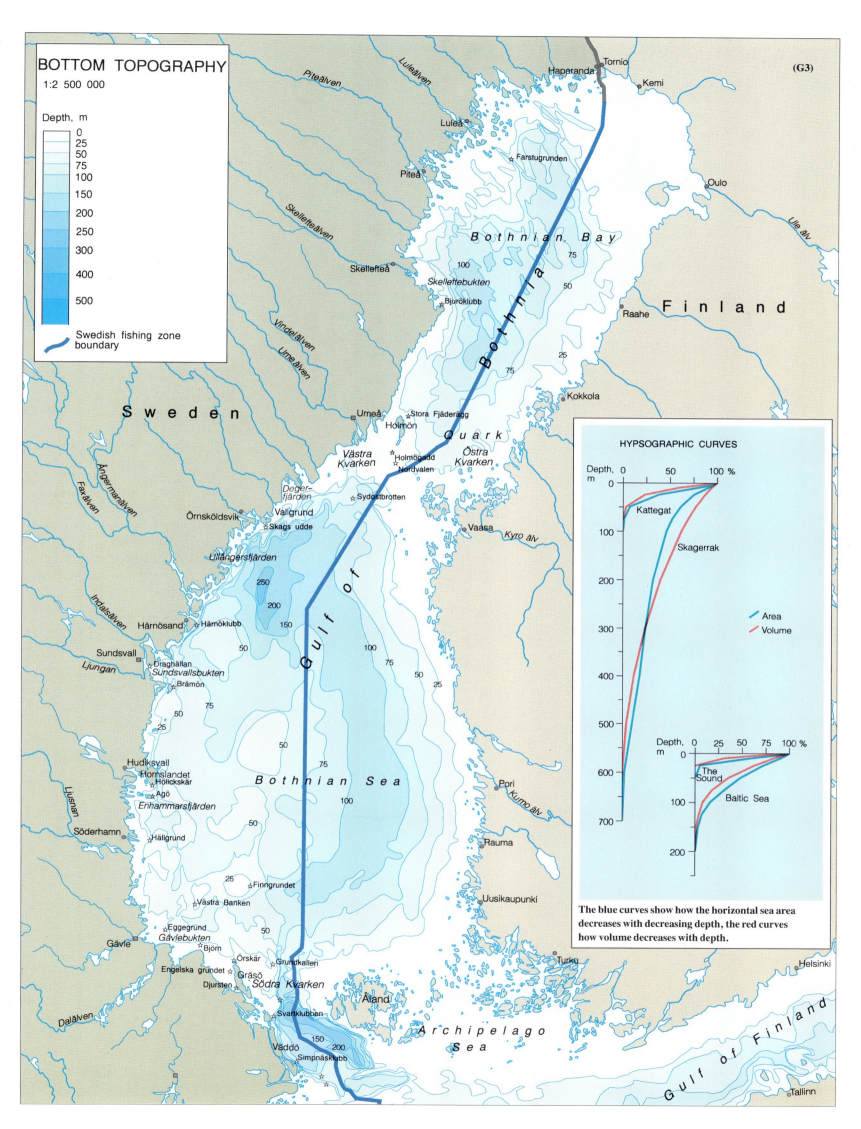

the Northern Åland Sea. They are separated from each other by the ridge between Söderarm and Lågskär. The greatest depth (301 m) is in the Northern Åland Sea. The Northern Åland Sea is separated from the Bothnian Sea by the Southern Quark (Södra Kvarken), through which a deep channel of about 100 m leads into the Bothnian Sea. The Archipelago Sea, between Åland and the Finnish mainland, is very shallow but three 30–40 m deep channels lead through it from the Baltic Proper to the Bothnian Sea.

To the north, the Bothnian Sea is bordered by the Quark which has a sill depth of 25 m. The greatest depth in the Bothnian Sea, the Ulvö Deep (Ulvödjupet), is off Ulvön, which the chart reports to have a depth of 293 m.

The Bothnian Bay is a relatively shallow sea, but the bottom is very uneven. The greatest depth is 148 m and is to the southeast of Luleå.

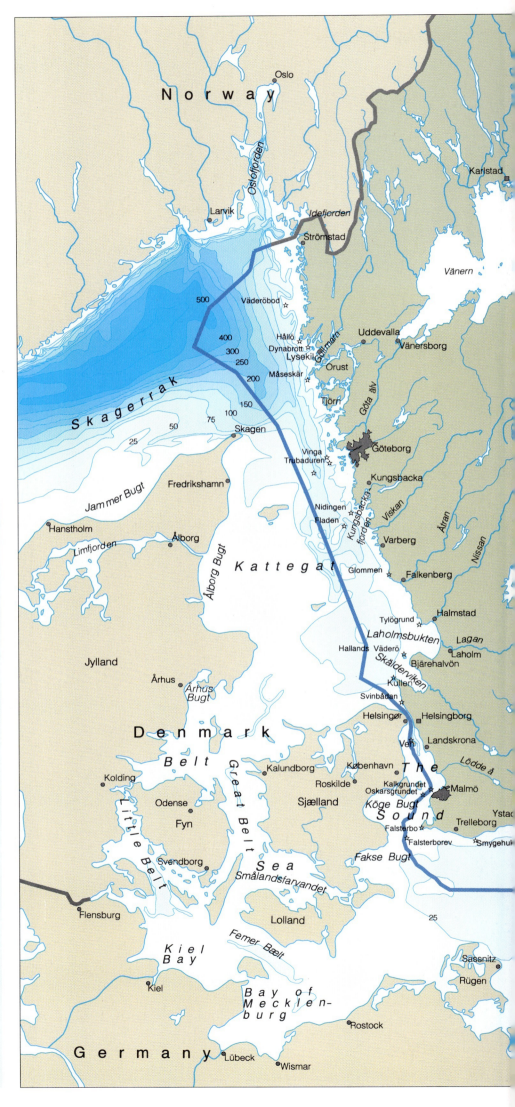

SEA AREAS

Name	Area (km²)	Volume (km³)	Mean-depth (m)	Max. depth (m)
BALTIC SEA	377,400	21,200	56	451
Gulf of Bothnia	115,640	6,380	55	293
Bothnian Bay	32,920	1,420	43	148
Quark	4,590	90	20	65
Bothnian Sea	63,190	4,300	68	293
Åland Sea	4,770	400	83	301
Archipelago Sea	10,170	180	18	123
Gulf of Finland	29,510	1,110	37	115
Gulf of Riga	18,560	410	22	56
Baltic Proper	213,690	13,300	62	459
Northern Baltic Proper	40,510	2,960	73	459
Central Baltic Proper	70,750	5,410	76	249
Southern Baltic Proper	96,650	4,640	48	131
Bay of Gdansk	5,780	290	50	117
BELT SEA	18,740	270	14	81
Bay of Mecklenburg	4,680	80	16	31
Kiel bay	3,400	60	16	36
Little Belt	2,400	30	14	81
Great Belt	8,260	100	12	47
THE SOUND	2,330	30	11	50
KATTEGAT	22,090	510	23	124
SKAGERRAK	31,530	5,490	174	700

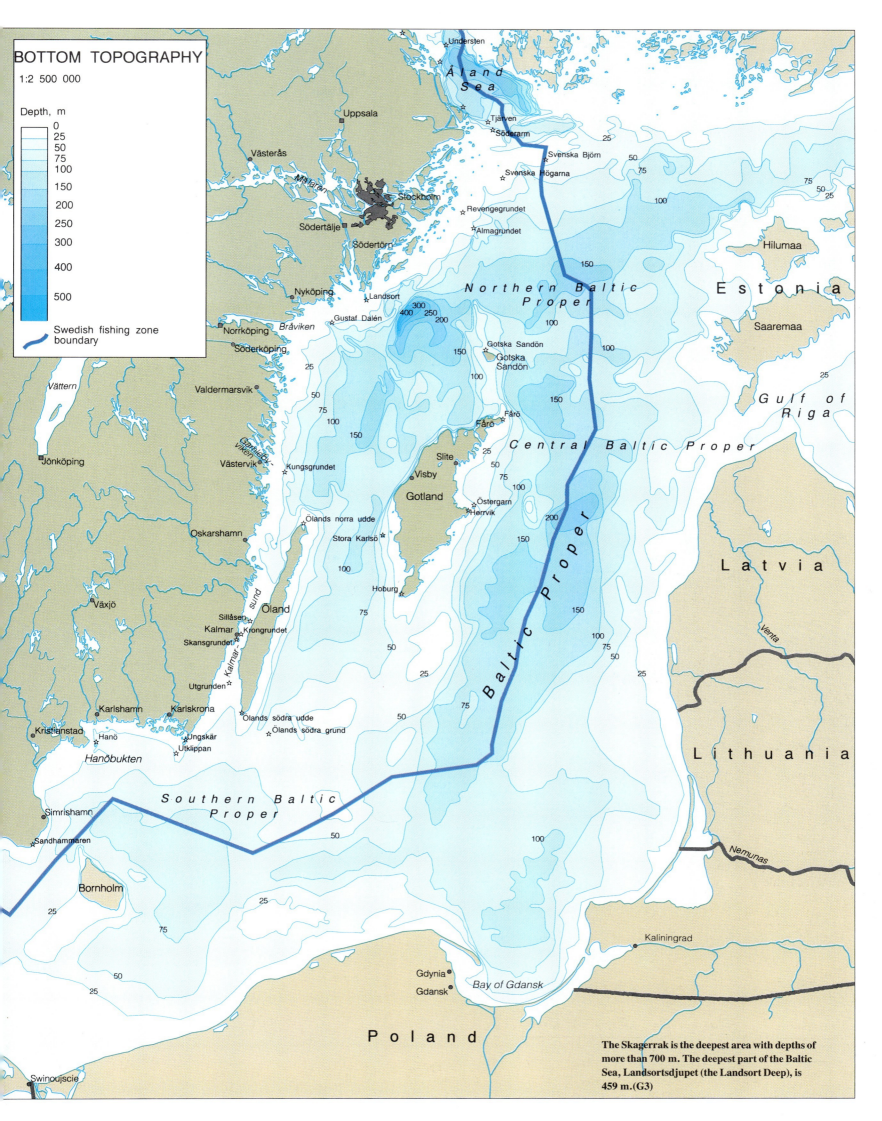

Development of the Seas

In a geological perspective, the present extent and bottom conditions of the marine areas around Sweden are of very young age. Most of the sediments were deposited during or after the last glacial period, and the extent of the sea has been formed through a complicated interaction between the movements of the Earth's crust and the surface of the sea (isostatic and eustatic changes, respectively). In this connection, however, the condition of the bedrock has been of the greatest importance and it has a long history.

Pre-Quaternary

The crystalline basement in southern Norway, Sweden and Finland is called the Baltic shield and makes up part of the east European platform. This was where the once massive but today eroded mountain chains we call the Svecofennids and the Karelids were found. The oldest rocks in the innermost part of the Bothnian Bay have an age of 2.6–2.8 billion years, whereas most of the basement in the rest of Sweden is between 1 and 2 billion years old. In the northwest, the Baltic shield is bordered by the Scandinavian mountain chain and to the southwest by the Tornquist Zone which extends from the Skagerrak to the Kattegat and the southern Baltic all the way down to the Black Sea.

To the southwest of the Tornquist Zone the basement sinks in steps down to depths of several thousand metres below Denmark and the southwest part of the Baltic States, where it is overlaid by massive younger sedimentary rocks. In the rest of the Baltic, the surface of the basement slopes relatively uniformly towards the southeast. From a depth of about 200 m along a line to the south of Blekinge, through Öland, Gotska Sandön and further in towards the Gulf of Finland, the basement is overlaid by, towards the southeast, increasingly massive sedimentary rocks. At the Polish coast, the basement is at a depth of more than 3,000 m.

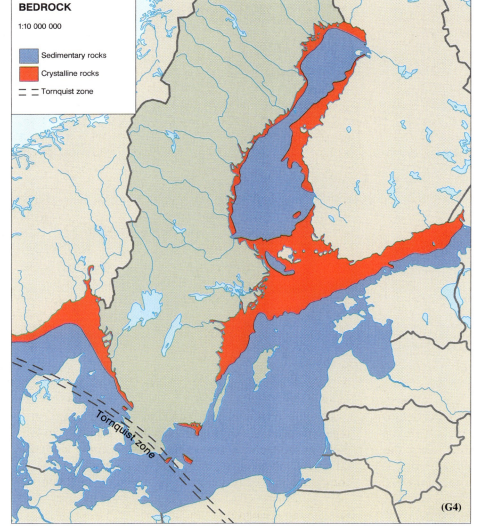
(G4)

Despite its relatively level surface, we speak of the sub-Cambrian peneplain, the basement shield is crisscrossed by several lines of fissures and faults. Similar faults in the basement rock have, for example, given rise to the Landsort Deep, the deepest part of the Baltic. In the same way, faults run along the western extremity of the Bothnian Sea and the Bothnian Bay, where the basement has a depth of about 500 m before slowly increasing up towards the Finnish coast to the east. About 1.3 billion years ago weathering products, e.g. sand, were deposited in the cracks and faults of the basement shield. With time, these deposits, often reddish in colour, were converted into what we today call Jotnian sandstone.

During the subsequent hundreds of millions of years, the Baltic depression was a shallow bay for long periods, in which sedimentary rocks were deposited. Thus, for example, fossils from Gotland show that during the Silurian, the Baltic was a sea with water that was so warm that coral reefs could be formed. At the end of the Tertiary, the depression became part of a land area characterized by rivers and lakes. It was not until the Quaternary, with its shifts between glacial periods and warmer periods, that the Baltic depression developed towards its present shape.

Quaternary

In the seas around Sweden, we can only identify with certainty the deposits from the last two glacial periods and the warmer interglacial period, as well as deposits from our own time (the postglacial). Traces of the oldest of these two glacial periods (Saale) and the subsequent interglacial period (Eem) can be found in several of the large shallow areas of the Kattegat. In other respects, the marine areas around Sweden are covered by deposits from the latest glacial period (Weichsel) and the interglacial period (the Holocene) we are today living in.

The Weichsel ice had its maximum extent more than 20,000 years ago when the border of the ice was in northern Germany and the North Sea. The British Isles were also covered by ice. About 14,000 years ago,

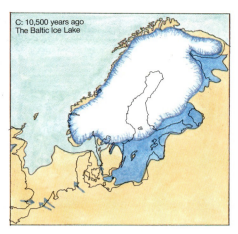

A. The latest glaciation—the Weichel—had its greatest extent in northern Europe at this time. At the same time the surface of the oceans was at its lowest level, about 130 m lower than today.

B. The ice margin retired and the Skagerrak, Kattegat and parts of Skåne became ice-free. The surface of the oceans was at that time about 100 m lower than today.

C. During a period with colder climate, the melting of the ice ceased and terminal moraines were formed along the edge of the ice in southern Norway, central Sweden and Finland. The Baltic Ice Lake, dammed up by the ice, probably drained through the Sound to the oceans.

D. When the margin of the ice retired from the northern point of Billingen and the water level in the Baltic Ice Lake sank to that of the oceans (about 60 m lower than today), brackish water was able to enter the Baltic Basin through the Central Swedish Sounds.

E. The rapid isostatic upheaval caused the sill of the Baltic Basin in central Sweden to rise above the present level of the oceans and the Baltic returned to a freshwater state. Initially, it drained through Lake Vänern but later the outflow changed to the Great Belt.

F. The ocean, the level of which still rose has penetrated through the Danish Sound and the Baltic Sea is again connected with the Skagerrak/Kattegat.

as a result of climatic improvements, the border of the ice had withdrawn to Skåne in southern Sweden, and about 10,000 years ago it had receded to the area of Lake Vänern and Stockholm. The final remnants of the ice sheet disappeared about 9,000 years ago.

Land Uplift and the Highest Shoreline

The weight of the massive glaciers caused the crust of the Earth to be compressed. Following the melting of the ice, the Earth's crust started to rise again but more slowly than the surface of the world's seas on account of all the meltwater entering the sea. This resulted in large parts of Sweden initially becoming inundated before again emerging from the sea and becoming land. The border where the highest positioned shoreline marks can be found today is called the highest shoreline. From this position, the coast-line has been shifted on account of isostatic uplift – land upheaval – to its present position. Evidence of this can be found in former raised beaches, deltas and areas far inland that have been washed bare by water. The highest shoreline rises from about the present sea level in Skåne up to 285 m above sea level at the coast of Ångermanland and then further to the north decreases to a height of 180 m in Tornedalen.

The developments in the Baltic depression were complicated. Its first stage, the Baltic Ice Lake, was replaced by the Yoldia Sea with connections to the ocean, but later was again separated into a lake, the Ancylus Lake, for a period before the connections were again re-opened leading to the saline Littorina Sea, which was followed by the brackish Baltic Sea.

The efforts of the Earth's crust to regain its original position are still ongoing but today with a maximum land uplift of 8 mm per year along the coastal districts of Ångermanland. This is only one-tenth of the initial rate. In the area around Stockholm, land uplift is about 4 mm per year whereas in southern Skåne, there is a land subsidence of 1 mm per year. However, these are not true movements since in reality land uplift (isostatic movement) is greater but is compensated partly or entirely by a simultaneous increase in the level of the sea (eustatic movement). Evidence of the latter can be found in submerged forests in, e.g., Hanö Bay (Hanöbukten), and the massive erosion to which the south coast is exposed.

Land uplift leads to certain consequences for Sweden. The total Swedish land area is increasing, whereas the water volume in the Baltic is decreasing by about 10 million m^3 annually. Land uplift in areas close to the shoreline has resulted in boathouses and jetties which were built a century ago today finding themselves on dry land; also both the groundwater level as well as the erosion base in water courses have decreased at the same time as sea bays are replaced by clayey plains.

Bottoms and Sediments

Most of the loose sediments on the continental shelf are the result of much the same geological processes as those active on land. The deposits and many of their morphological shapes originate almost exclusively from the latest glacial period. The ice-cover reached a thickness of several kilometres and extended over large parts of northern Europe, including the sea areas around Sweden. As a result of the erosion of the ice sheet (land ice), large parts of the soil layers which had been deposited during earlier glaciations became re-worked and re-deposited together with material that the ice had broken off from bedrock outcrops. The sedimentary bedrock was the most susceptible to erosion and therefore only remained intact in protected positions.

The soil and rock material taken up and worked by the ice was gradually deposited when the ice receded. If the material was directly deposited by the ice, this gave rise to a soil type with a mixture of all grain sizes, called till. This is the most common soil type in Sweden. If the material was sorted by the meltwater from the ice, then glaciofluvial deposits were formed. In tunnels beneath and in the ice, and where these tunnels opened at the edge of the ice, eskers were formed, whereas deltas were formed by the glacial rivers beyond the edge of the ice where they flowed out into open water. The finest particles, the suspended load of silt and clay, were deposited at greater distances from the ice and formed, for example, varved clays. During occasional periods when the ice-margin was stationary, formation took place of terminal moraines and ice-marginal deltas. Examples of this can be seen, for example, near Varberg, where the coastal moraines of Halland form distinctive features of the landscape and the neighbouring seabeds off the coasts. Eskers are found in many places. Perhaps the most renowned is the northernmost land offshoot of the Uppsala esker, Billudden, and its continuation northwards in the Bothnian Sea.

Sediments

Only a very small part of the seabed of the seas around Sweden has been systematically mapped. Of Sweden's part of the continental shelf extending over more than 160,000 km², only 10% has been systematically mapped as regards the distribution of sediments. The map on the right, therefore, is based on other published marine-geological works, frequently of varying character and with varying degree of cover, complemented with seabed information provided by the navigational charts. The division of sediments used is based on both grain size distribution and the genesis of the sediment. The map is intended to provide a schematic picture of the geological characters of the various bottom areas.

CLAY BOTTOMS

Clay has been sub-divided into glacial (deposited during the Ice-Ages) and post-glacial clay on account of the major importance of these clays in identifying bottom dynamic conditions. Postglacial clay and silt, often with a concentration of organic material which allows the sediment to be characterized as gyttja clay, is fine-grained sediment which has been deposited after the recession of the latest ice sheet and which is characteristic of bottom areas where sedimentation is still ongoing (deposition bottoms). Gyttja clay, often grey-green in colour, usually fills to some extent, but with increasing amounts at greater depths, the depressions in the seabed and in this way levels out the bottom profile. This is the situation in, e.g., the Skagerrak, with its great depths along the Norwegian Channel.

On the other hand, in the Kattegat the largest present deposition of fine-grained material is taking place along the western edge of the Deep Channel where mainly material carried by currents around the northern point of Jutland is deposited down the slopes. Even further to the south, in the deep channels to the west of Fladen and Lilla Middelgrund, there are a number of places with different deposition patterns. Here, the clay is deposited asymmetrically, with the greatest amounts along one edge of the bottom of the winding deep channel, probably depending on the longitudinal direction of bottom currents.

In the Sound and the Belts, the cross-sectional area of the water masses is small and the current velocities are consequently large. In such places, fine-grained sediment is deposited to a smaller extent and only in isolated deep depressions. The Baltic Proper, on the other hand, is characterized by several large, level, sedimentation basins, e.g., the Arkona Basin, the Bornholm Basin and the Gotland Deep. In such places, the sedimentation rate today is between 0.5 and 1.5 mm/year in the central parts of the deep areas and decreases to zero in peripheral parts. Values of the same magnitude have also been calculated for the large sedimentation basins of the Bothnian Sea and Bothnian Bay. In some of the protected deep holes, which are also found close to the coast and in shallower waters, there may be considerably larger sedimentation rates locally.

The glacial clay, which was deposited as the distal sediment of the glacial rivers in connection with the melting of the Scandinavian ice sheet, has particles that are just as small as those in postglacial clay but with a considerably lower content of organic matter. In contrast to postglacial clay, in places where glacial clay makes up the bottom of the sea, it indicates those areas which are exposed to erosion or transport of material as a result of the influence of waves or currents. Thus, areas with glacial clay can be found in exposed positions in shallower waters off the Swedish west coast. In the Baltic, the glacial clay, usually red-brown and often with clear annual varves, is found on slopes down towards the larger sedimentation basins and also in exposed positions and in shallow areas along the coasts, e.g., in the Hanö Bay, along the east coast of Öland, or in more offshore areas of the Svealand archipelago. In the Bothnian Sea, the conditions are similar, whereas exposed glacial clay becomes less common in the Bothnian Bay where it is generally covered by younger sediment even far up on the slopes.

This outline map is based on several marine geological publications and reports. A more detailed description is given in the text on the left. (G5)

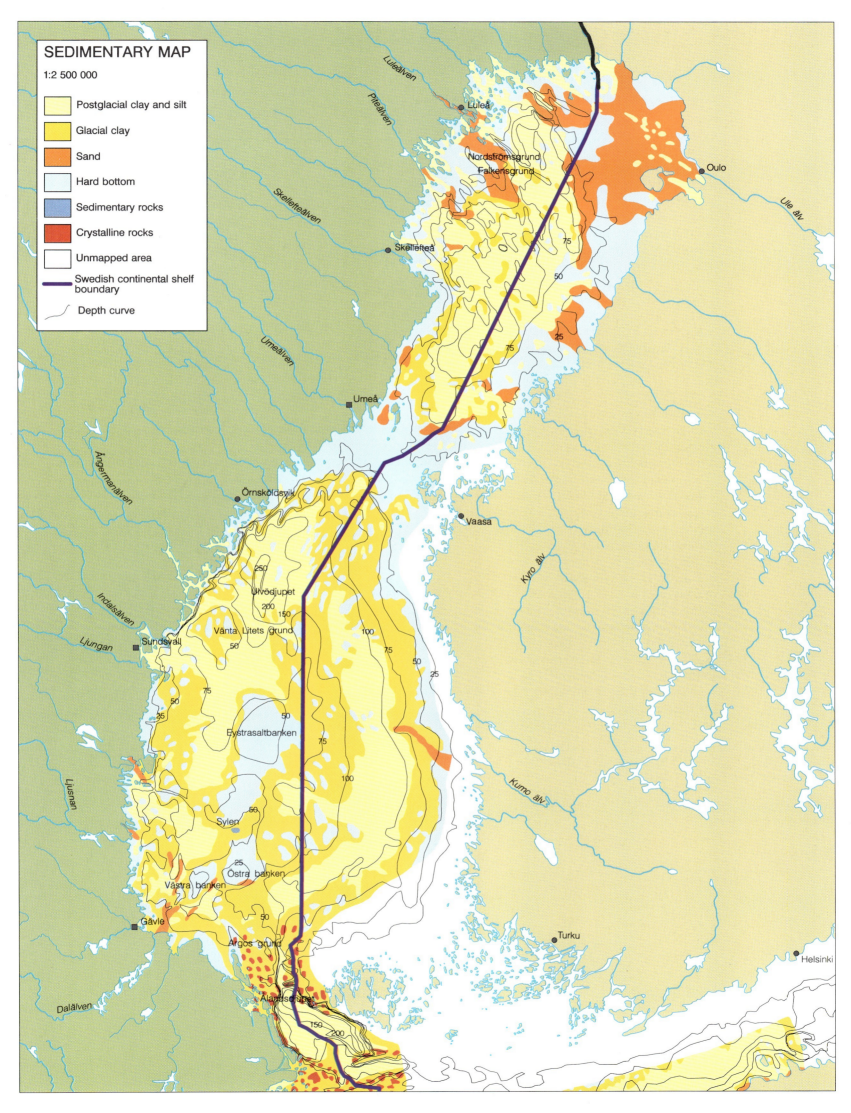

SAND BOTTOMS

Sand bottoms are usually an indication of active transport and reworking of material. Sometimes, clear bed forms such as ripples or sand waves may illustrate the prevailing direction of transport through its gently sloping stoss side and steep lee side. If sand from a shallow area is transported out over a slope and is accumulated at greater depths with less influence of waves and currents, then it may result in sand accumulations of considerable magnitude. Primary, glacial sand accumulations, e.g., eskers, directly on the surface of the seabed are less common. They are generally reworked as a result of wave and current activity, or are at greater depths, mainly covered by fine-grained sediments.

HARD BOTTOMS

The description "hard bottom", which is used on the map showing sediment distribution, covers a wide range which includes the till deposits of the inland ice as well as the different residual sediments such as gravel, stones and boulders, which may be the result of wave and current erosion and the transport away of fine-grained material to deeper sedimentation areas. In several places, glacial deposits may assume considerable thickness. Thus, for example, several of the "banks" in the Kattegat are largely composed of till, frequently together with older glacial or interglacial fine-grained sediments which were later affected by the land ice through, e.g., folding.

The Swedish continental shelf covers about 160,000 km², corresponding to about 40% of Sweden's land area. The Geological Survey of Sweden had mapped about 10% of the shelf by 1990.(G6)

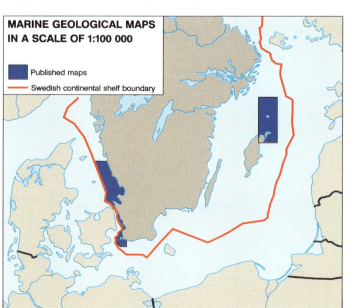

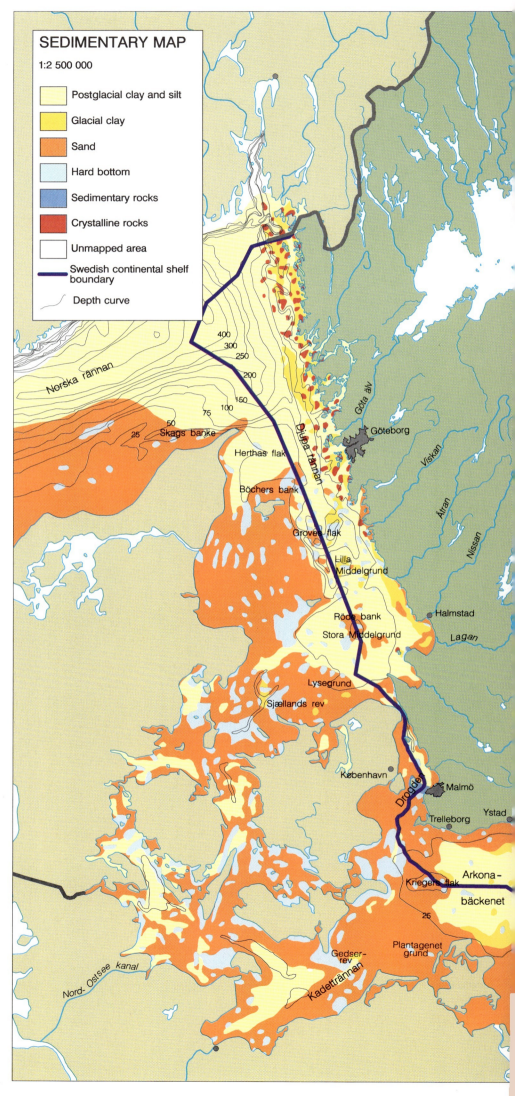

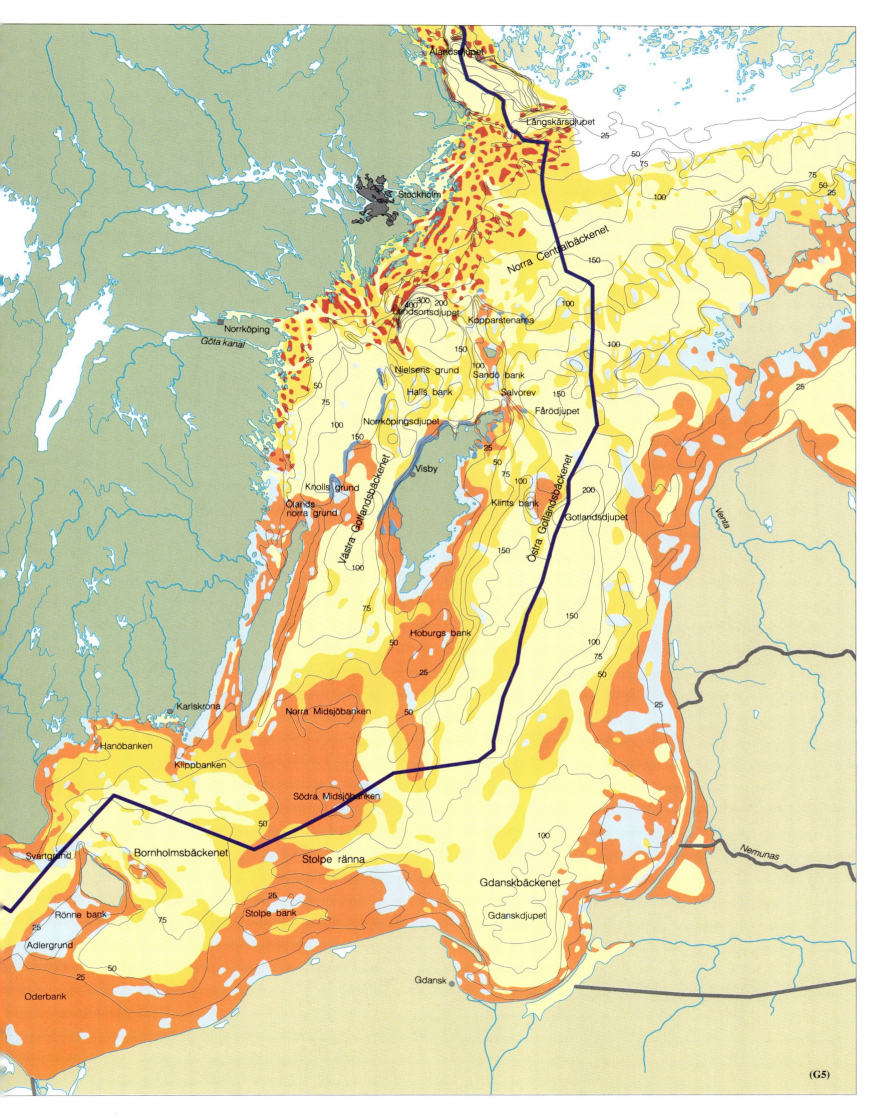

Postglaciala bildningar
Postglacial deposits

Lera, gyttjelera och lergyttja
Clay, gyttja clay and clayey gyttja

Finmo och mjäla
Silt

Grovmo
Fine sand

Sand och grus (huvudsakligen sand)
Sand and gravel (mainly sand)

Glaciala bildningar
Glacial deposits

Glacial lera
Glacial clay

Glacial mo och mjäla
Glacial fine sand and silt

Isälvsavlagring i allmänhet
Glaciofluvial deposit, unspecified

Morän
Till

Äldre glaciala/interglaciala bildningar
Older glacial/interglacial deposits

Huvudsakligen lera/mo, vanligen veckad
Mainly clay/silt, normally folded

Berggrund
Bedrock

Sedimentär berggrund
Sedimentary bedrock

Kristallin berggrund
Crystalline bedrock

Block på annan jordart än morän
Boulders on other deposit than till

Tunt torvlager
Thin peat layer

Finmo, mjäla el. lera under tunt lager av annan, med färg markerad, jordart
Silt or clay below thin layer of other sediment which is marked in colour

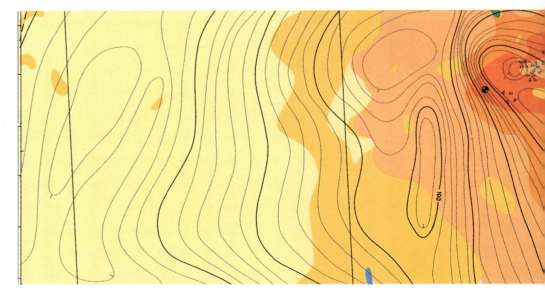

Part of the Marine Geological Map of Gotska Sandön.

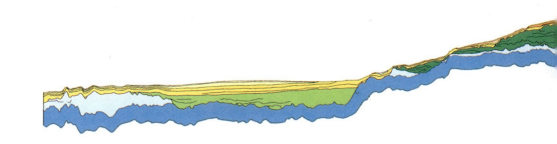

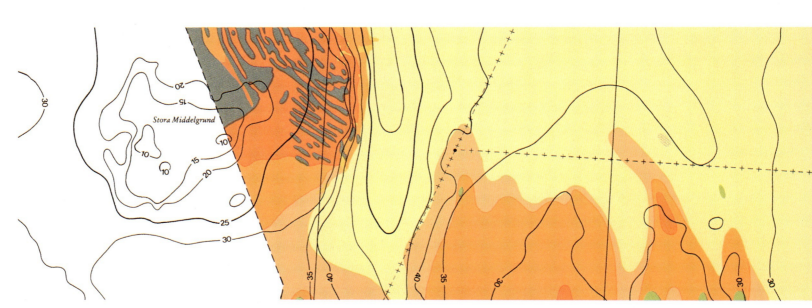

Part of the Marine Geological Map of Stora Middelgrund-Halmstad.

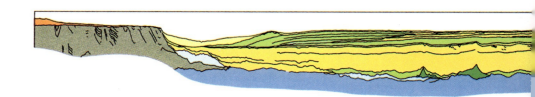

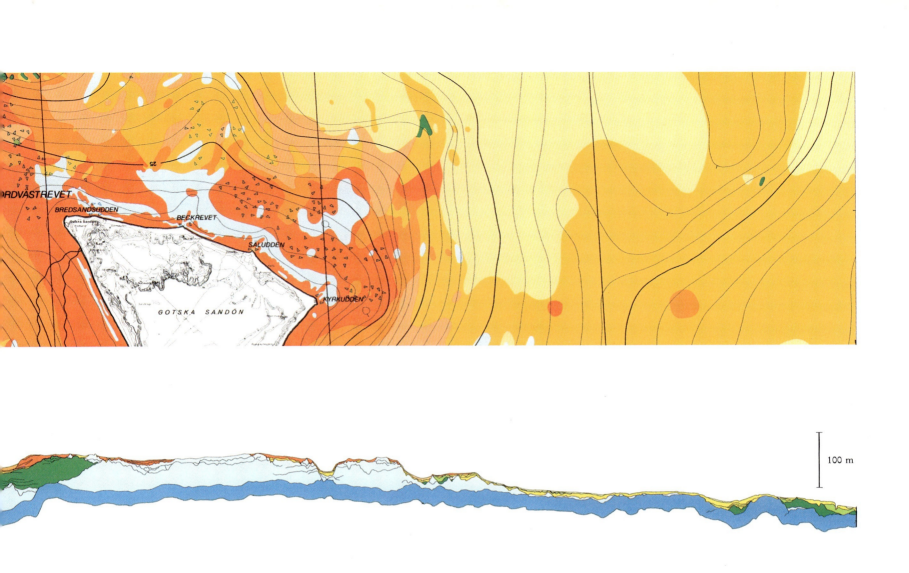
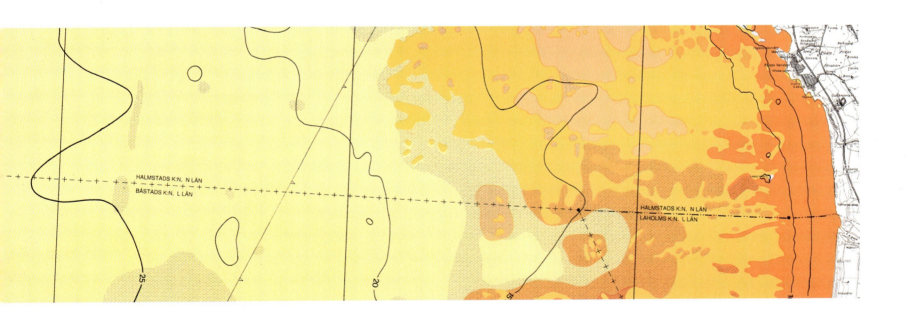
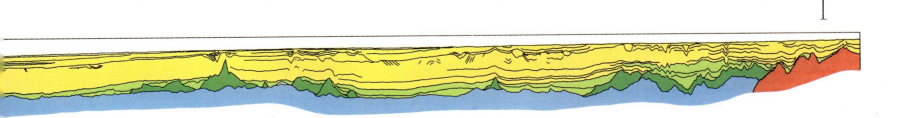

Erosion and Deposition

The bottom sediment is exposed to varying degrees of influence depending on the hydrographical conditions prevailing in the area. If the conditions change, then the bottom sediment also changes character. The influence of waves and currents causes erosion, transport and deposition of bottom material and thus we distinguish between these three types of bottom. The degree of influence depends mainly on water depth, current velocity and the condition of the bottom material.

The shoreline of the Bothnian Sea and the Baltic Proper, to the north of a line slightly to the south of Gotland and northern Skåne, is the lowest since the glacial period. This implies that bottom areas on the seaward side are to a minor extent exposed to the reworking and wave-wash which is so typical of the coastal zone and the formerly submerged areas of land. Along the coast of Skåne and the coast of Halland up to Varberg, however, the shoreline was earlier 15–20 m lower. These areas, today submerged, were thus earlier exposed to strong erosion and reworking.

Within the Baltic today, about half of the seabed is exposed to erosion or transport. The reworking of sediment, which occurs as a result of this, is estimated to be six times larger than the supply of sediment with the rivers. The mean value of the annual sediment increment within the deposition areas of the Baltic amounts to between 0.5 and 2 mm.

The Kattegat is characterized by few deposition areas, whereas the opposite applies to the Skagerrak. Apart from rivers and the result of reworking of their own shallow areas, both the Skagerrak and Kattegat receive suspended material which is supplied in the water masses which flow in from neighbouring areas, the North Sea and the Baltic. Of the ca. 25 million tonnes of suspended material supplied to the southern North Sea annually, about 17 million tonnes are estimated to be taken further with the Jutland current to the Skagerrak and Kattegat. Altogether, it is estimated that 30 million tonnes are sedimented in the Skagerrak annually. The average sedimentation rate is 2 mm/year, but extreme rates of more than 100 mm/year have also been observed in the western part of the Deep Channel.

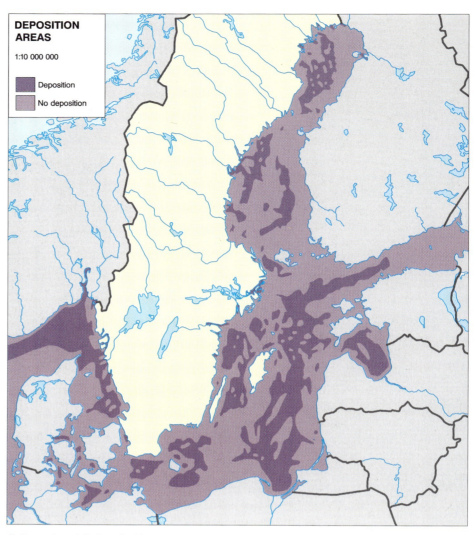

Sediment is mainly deposited in the deep parts of the sea, which are not so exposed to waves and currents.(G7)

What the Shelf Sediment Reveals

The different sediments on the seabed contain evidence of the bottom dynamic processes that have been predominant. The fine-grained sediments from areas with undisturbed sedimentation over long periods which have resulted in the sediment bed successively growing in thickness are of particular interest. These fine-grained sediments make up an historical archive in which the sea environment and its changes can be followed backwards in time by studying increasingly deeper layers of sediment. Variations in the microfossil content and the mineralogical and chemical compositions between different layers of sediment illustrate, for example, changes in climate, water temperature, oxygen availability, salinity, nutrient supply and areas in which the sediment originated.

OXYGEN AND HYDROGEN SULPHIDE

A large part of the present environmental problem in the seas surrounding Sweden is associated with a deficient availability of oxygen. Oxygen deficiency can be identified in the sediment when the normal, oxidized, light grey-green surface of the sediment becomes dark grey or black as a result of iron compounds being reduced to ferro-monosulphides.

Nonetheless, it is important to remember that oxygen deficiency and hydrogen sulphide are not new events in the Baltic. The sediment illustrates that the deep areas have suffered from these problems long before mankind could influence the situation. The risk of oxygen deficiency is, quite simply, unavoidable as a result of poor water exchange in certain deep areas, which in turn is an effect of, e.g., the shallow sills of the Baltic (e.g., in the Sound) which only periodically allow inflows of heavy oxygen-rich bottom water. However, during recent years the situation has become more acute owing to the increased supply of nutrients.

HEAVY METALS

The recent (the latest to be deposited) fine-grained sediments have attracted particularly great attention during past decades for the possibilities they offer in environmental monitoring. Awareness that heavy metals and persistent organic compounds originating from different forms of discharges into lakes, rivers and the sea are stored in the bottom sediment has resulted in them becoming an important implement in the work of documenting and monitoring anthropogenic influence on the environment. In particular, sediment from calm sedimentation environments along the coasts and in the estuaries, which geologically have a high sedimentation rate, has been found useful. During the last two decades in Sweden, as elsewhere in the world, sediment cores have been successfully used to document the increasing supply of these substances which are hazardous to the environment.

This is a result of the fact that many elements in the sea, e.g., heavy metals, are taken up by organisms and/or bound by processes such as adsorption, precipitation, ion exchange, complex-formation, etc. to organic and inorganic particles, which will gradually sink through the water mass and become incorporated in the seabed. The content of heavy metals in the sediment, however, is controlled not only by these processes but also by factors such as the hydrography and bottom dynamic conditions, the ambient pH (acidity) and redox situation (access to oxygen). In addition, the physical status and

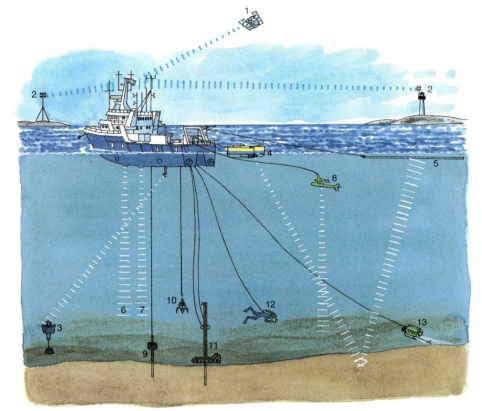

Mapping of the seabed

POSITIONING SYSTEM
1. Satellite positioning system
2. Radio positioning system
3. Hydroacoustic positioning system

HYDROACOUSTIC SURVEY METHODS
4. Seismic sound source
5. Hydrophone
6. Sub-bottom profiler
7. Echo sounder
8. Side scan sonar

SEDIMENT SAMPLING METHODS
9. Gravity corer, piston corer
10. Grab sampler
11. Vibrohammar corer

OBSERVATION METHODS
12. Diver with UW-camera
13. Underwater video camera

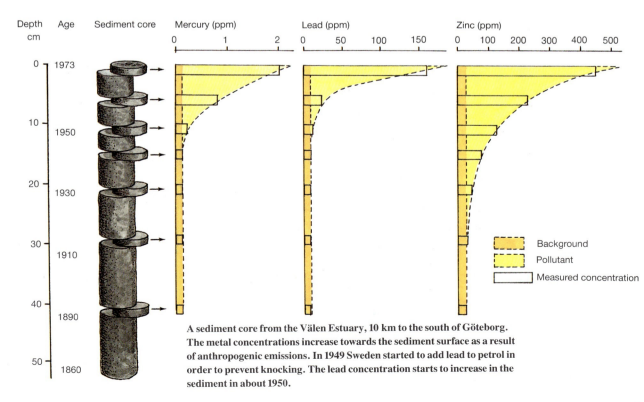

A sediment core from the Välen Estuary, 10 km to the south of Göteborg. The metal concentrations increase towards the sediment surface as a result of anthropogenic emissions. In 1949 Sweden started to add lead to petrol in order to prevent knocking. The lead concentration starts to increase in the sediment in about 1950.

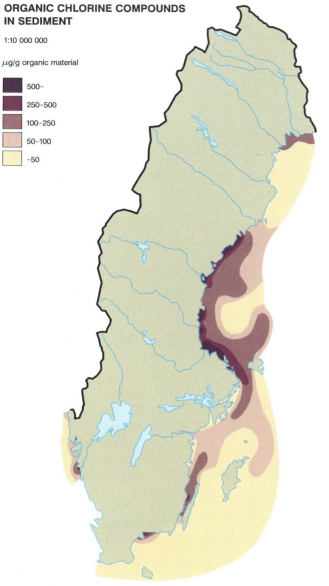

ORGANIC CHLORINE COMPOUNDS IN SEDIMENT
1:10 000 000

μg/g organic material
- 500-
- 250–500
- 100–250
- 50–100
- -50

The content of total extractable organically bound chlorine (EOCL) is strongly increased in the surface sediment in sea areas off pulp mills producing, bleached pulp. The map is based on data collected during 1984–1990. (G8)

Coastal water is used both as a recreational resource and also as an industrial resource. Motor cruisers and the ASSI Kraftliner factory off Piteå.

cisive importance to varying degrees.

It is particularly the organic material, which sooner or later becomes deposited on the seabed, which has an extremely great capacity to concentrate and bind heavy metals. This implies that a high metal concentration in one part of a sea area (recipient) may be caused only by a high organic content in the sediment which, in turn, often reflects the sedimentation environment.

The Gradient method, used in some areas along the Swedish west coast during the 1970's and 1980's, identifies Idefjorden, Stenungsund and, pre-1974, Välen, as the most polluted coastal areas in the Skagerrak/Kattegat. As regards mercury and cadmium, the Göteborg archipelago should also be included in this extremely seriously loaded group of marine areas. Byfjorden and Askimsviken are two other relatively heavily loaded coastal sectors.

RADIOACTIVE ISOTOPES

Large emissions of radioactive isotopes have taken place in connection with the atmospheric nuclear tests conducted during 1961 and 1963 and after the reactor accident at Chernobyl in 1986. In the same way as other metals, the radioactive isotopes become accumulated in the seabed. Activity differences in sediment cores can, for example, be used to date these events.

OIL AND CHLORINE

The different petroleum compounds entering the sea as a result of combustion of fossil fuels, shipwrecks, etc., meet different fates. Some are decomposed relatively quickly by microorganisms or sunshine, whereas others, e.g., the poisonous polycyclic aromatic compounds (PAH), are more resistant. The latter may remain for one or more decades in the sediment and can be found mainly off residential areas and refineries.

The dominating source of discharges of chlorinated organo-compounds in the sea areas around Sweden today is the pulp industry. The dispersal of the chlorinated substances has been surveyed by analysis of the contents of extractable organic bound chlorine (EOCL) in the sediment. The results show strongly increased concentrations in the sediment off the pulp factories. The concentrations are also noticeably increased further out in the open Bothnian Sea.

Microfossils

When planktonic and benthic organisms die and come to rest on/in the bottom they are consumed by bacteria and animals. However, many types of microorganisms, just like snails and mussels, have shells which are preserved. Depending on the type of organisms, these shells may consist of different types of material. Calcite, silica, cellulose and other closely-related organic substances are among the most common types of shell material which can be preserved as fossils.

Studies of the microfauna and microflora in the uppermost sediment layer offer a picture of how the physical and chemical conditions at that place characterize species composition. If studies are required of how salinity, water depth, climate or other environmental conditions have varied or changed through the years, this can be done by studying the microfossil content in the deeper layers of the sediment. If a sediment core is sliced and the species composition is gradually investigated downwards in the core, i.e., backwards in time, it is possible to identify the environmental conditions prevailing at that place over several years, several centuries, millenia or billions of years in the past, depending on the length of the core and the sedimentation rate prevailing at that place. The most common types of microfossils studied by geologists are *microalgae*, e.g., diatoms, dinoflagellates and coccolites, *amoeba with shells* (foraminifers and radiolaria), *ostracods, pteropods* as well as *pollen* and *spores* from terrestrial plants which have been transported out to sea from neighbouring areas of land. Representatives of all these types of microfossils are found in the quaternary sediments of the Kattegat and Skagerrak.

MICROFOSSILS IN THE KATTEGAT AND SKAGERRAK

Sediment cores from the Kattegat and Skagerrak provide information on how the sediment during the final phases of the glacial period was characterized by Arctic (cold) fauna and flora and how these later, after the end of the glacial period about 10,000 years ago, changed to organism communities requiring warmer conditions. New species become established and others disappear as a result of the environmental changes even long after the dramatic end of

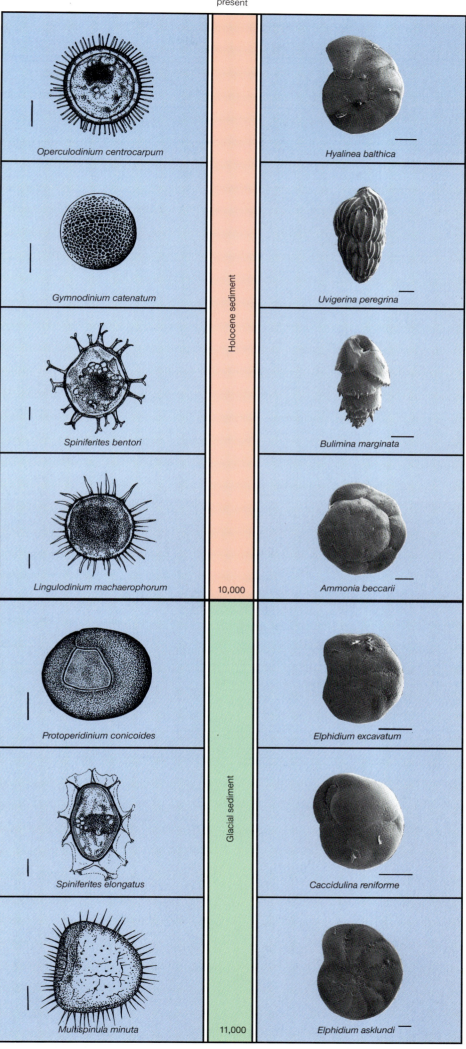

Some of the commonest dinoflagellate cysts and foraminifers found in the sediments of the Skagerrak and the Kattegat are shown here. Cysts 1, 3, 4, and 5 from the top are illustrated with cell contents which otherwise only occurs in living cells. Scale: 10 μm for dinoflagellate cysts and 100 μm for foraminifera.

the glacial period. Such changes may be caused by, e.g., lower depths of water caused by land uplift, climatic variations, changes in the patterns and velocities of marine currents, or by anthropogenic influence.

Here, we will concentrate on two groups of microfossils which are abundantly represented in the sediment of the Skagerrak and Kattegat, namely dinoflagellates, with organic cell walls, and foraminifers, with calcareous shells.

Dinoflagellates are usually grouped among the microalgae and make up a large and important group among our marine phytoplankton. Several species produce resting cells (cysts) which are preserved in the sediments. The motile cells and their resting cell stage are usually relatively similar in size and generally vary between about 0.02 and 0.15 mm.

The dinoflagellate flora in the Kattegat and Skagerrak during the final phase of the glacial period was characterized by species which today are found in Arctic waters. After the end of the glacial period, a flora became established which preferentially inhabits temperate waters and which largely corresponds to the one we find in the Skagerrak and Kattegat today.

The sediments deposited in the Kattegat during the period from 2,000 up to 300 years ago contain an exotic species, at least for our waters – called *Gymnodium catenunatum*.

The foraminifers are marine amoeba with shells which occur both as planktonic and as benthic (bottom-living) animals. The benthic types dominate in the Skagerrak and Kattegat, whereas the planktonic types are more typical of oceanic conditions. Foraminifers usually have a size of 0.1–1 mm, depending on species. The benthic foraminifers found in the sediment deposited in the Skagerrak and Kattegat during the final phases of the glacial period are typical of arctic environments. Today, the same species are found in the seabed around, e.g., Greenland. The arctic fauna are relatively poor in species and generally do not contain more than 5–10 abundant species. After the end of the glacial period, a more temperate fauna, with a larger number of species, became established along the Swedish west coast. However, their species composition varies strongly depending on the depth of water, salinity, nutrient availability, bottom material and oxygen conditions.

Coast and Shore

Some coastal areas are strongly influenced by the effect of the sea, whereas others have not developed any special coastal characteristics. The resistance of bedrock to weathering and erosion, and changes in the levels of the sea surface and the Earth's crust determine that most of the shaping of our coasts is the result of a very long development where the condition of the coast today is only a very minor part. The coastline today is largely a rather arbitrary line drawn between ancient forms of landscape by reflecting, in a geological perspective, a temporary level of the sea surface. This leads to the Swedish coasts being more diversified than areas with more uniform development. The age difference between the large and small forms of the bedrock and the covering layer of soil is also characteristic. Many of the major configurations of the coastal landscape can be traced for hundreds of millions of years in the past. The smooth, rounded, surface of the rocks has not been shaped by modern waves but by the inland ice of several tens of thousands of years ago. On the open coasts the waves may throw up shingle a few metres above sea level, but the level of the shoreline has been moved as far as 285 m over the last 9,000 years.

There is a wide range between the terms coast and shore. The shore is restricted to the immediate contact between land and water as well as with the winds, currents and movements of the waves which may influence the seabed and land on either side of this boundary. It may feel natural to include uplifted sea-exposed shorelines within the term coast, as well as areas which, as a result of local climate and vegetation, have a coastal character contrasting with that of the hinterland. The outer extremities of the coast can be placed at the limit where the largest waves start to influence the bottom. This is a dynamic borderline which is governed by the magnitude of the storm-waves and thus varies on different coasts. In unprotected parts of Swedish coasts, this limit is at about 30–50 m deep.

Shore Zone Dynamics

The waves transform the coast into a dynamically unique environment. The length, height and frequency of the waves determines their dynamic effect. As the water gets shallower towards the shore, the movement of the wave is slowed down by friction against the bottom. The velocity of the wave's movement and the length of the wave decrease but the frequency remains constant. The flow of water in contact with the bottom may cause the bottom sediment to be disturbed. As a result of the movement to-and-fro, ripples are formed on sandy bottoms.

When the wave length decreases, the waves become steeper, and when the waves break over a bottom due to decreasing water depth and form breakers, the limit for the shore zone has been reached. Under the breakers, the water velocity towards the land beneath the crest of the wave is greater than the velocity from land below the troughs of the waves. The result is that particles of sand and gravel become sorted and make up larger transport forms. Finer particles which easily become suspended are taken outwards, sand is taken inwards and forms break-point bars. Long waves, rolling in over shallow bottoms, form surf waves with crests that are gradually broken down. Around the Swedish coasts, the waves are so steep that we do not get any wide, coherent zone of surf for surf-boarding. Bars are also formed below the surf waves moving in towards the shore with their steepest slope on the landward side.

Wave movements end in the swash zone where the wave rushes onto the beach in a state of strong turbulence. During the flow back in the backwash, the turbulence is weaker. This leads to a sorting with coarser and more rounded material in the lower part of the swash zone where a step is formed, and with finer and flatter particles in the upper part of the swash zone. The upward movement of the swash may lead to the formation of a beach-berm which will be the steeper the coarser the material it consists of.

The breakers and the surf move water in towards land. The water forced up in this way flows out in temporary, narrow surface currents which may have very high speeds. These rip currents have caused many drowning accidents in large sandy bays. They frequently change position and consequently it is impossible to identify risk-free parts of the beach. However, there are greater risks in the neighbourhood of protruding rows of piles or jetties which may turn the current outwards.

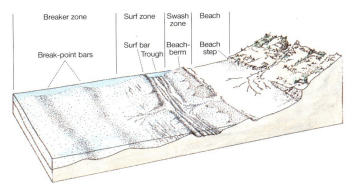

The dynamic zoning and characteristic bedforms of the beach.

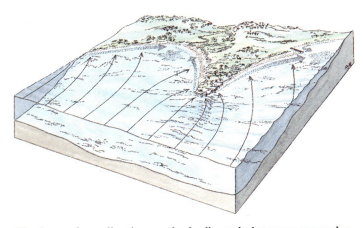

The change of wave direction over the shoaling seabed as a wave approaches the beach and the resulting currents and transport of sediment along the shore.

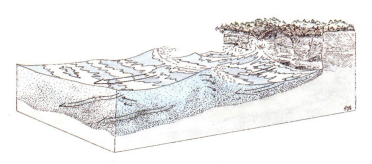

Wave erosion during storm and high water on a sandy beach with dunes.

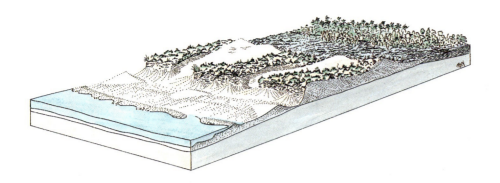

SANDY COASTS

On coasts with wide sandy beaches the wind may blow large amounts of sand in towards the land. On the naked backshore areas the wind will create delicate ripples and small dunes which migrate inwards or accumulate around seaweed or driftwood thrown up by the sea. Winds in excess of 7 m/s – a fairly fresh breeze – are necessary to set the sand in movement. The sand is caught by the vegetation further inland and forms irregular dunes. The main Swedish dune plants are lyme grass (*Elymus arenarius*) and marram grass (*Ammophila arenaria*). In places where the vegetation is damaged, the dunes may become eroded. The eroded sand will form new dunes further inland.

CLIFFED COASTS

Cliffed coasts have been created by the waves undermining a slope so that falls and landslides occur, after which the soil masses have been gradually washed away by the waves. A cliff can be formed both in rock and in loose material. As erosion of the cliff continues, the shallow bottom over which the waves approach will become increasingly wider and the wave energy will be distributed over an increasingly large area, whereupon erosion of the cliff will decrease. Stratified rocks such as limestone and sandstone, with a gently sloping layering, form particularly beautiful cliffs. Fractured crystalline bedrock, e.g., in a fault zone, will result in the cliffs and the beach having a very irregular form. Till cliffs generally contain widely differing types of soil, ranging from clay particles to large boulders. Consequently, the shape of till cliffs may vary widely. Large amounts of stones and boulders may form a natural armour at the base of the cliff and thus counteract continued erosion.

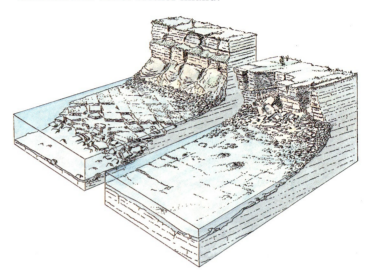

DELTA COASTS

Gravel, sand and finer materials which are transported by rivers and streams to the coast are mainly deposited off the mouth of the stream or river. Sand and coarser material is rapidly deposited when the current velocity decreases and forms a delta lobe with a mouth bar. When these bars grow, the current is divided into two channels, one on each side of the bar, after which two new bars are formed with a lagoon between the channels. Consequently, during its growth, the delta obtains a typically tree-like branched pattern. When the front of a delta is exposed to wave action, berms are formed which may cut off older inactive channels.

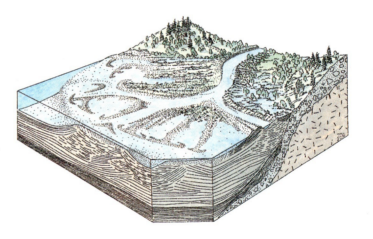

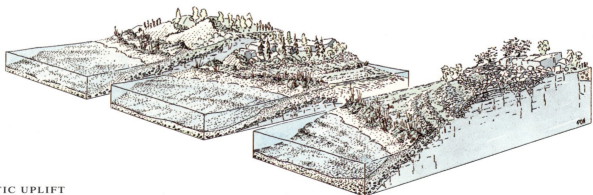

ISOSTATIC UPLIFT

Depending on bedrock, soil type, large-scale geomorphology, wave exposure, etc., isostatic uplift – or land elevation – may result in very different effects on coastal development. Higher prominent rocky heights are more or less washed clean and include features of frost-cracked boulders. Shingle beds are found in strongly wave-exposed slopes which, in less exposed positions, turn into gravel and sand terraces or into a saddle between two heights. Within active dunes with lyme grass (*Elymus arenarius*) we can find older dunes, more levelled-out, and frequently covered with bearberries (*Arctostaphylos uva ursi*) and crowberries (*Empetrum* sp.). Examples can also be found of renewed wind activity where the plant cover has been damaged and where sand has blown into forest hinterland behind the shore.

COASTAL REGIONS

1:5 000 000

- 🟩 Part of larger peneplain
- Limited peneplain
- 🟫 Hilly landscape
- ╱ Joint tectonics
- ⊥⊥⊥⊥ Major fault
- 3 Annual land uplift in mm
- 🟦 Coasts formed in sedimentary rocks or in soils on sedimentary bedrock
- Highest shoreline

Map labels:
- The archipelago coast of the Bothnian Bay
- The coast of the Northern Quark
- The High Coast
- The coastal plain of the Bothnian Sea
- The Southern Quark rocky barrier coast
- The Stockholm archipelago
- The Södermanland archipelago
- The Östergötland archipelago
- Kalmarsund, Öland and Gotland
- The Blekinge archipelago
- The open coasts of Skåne
- The lowland coast of the Kattegat
- The archipelago coast of the Skagerrak

(G9)

Coastal Regions

The coasts of the Bothnian Bay and Bothnian Sea are formed of pre-Cambrian peneplain with varyingly dominating features of joint tectonics which developed later into fissures and pre-glacial, today partly submerged, fluvial morphology, which extends out into the archipelagos. The high, hilly relief of the High Coast diverges from this pattern.

The same peneplain with fissure valleys dominates along the Baltic coast. The rocky hillocks of Blekinge, the cliffs and coastal plains of Skåne and the cuesta coasts of Öland and Gotland diverge from this pattern.

On the west coast, the Kattegat coastal plain and the archipelago of the coastline along the Skagerrak are most prominent.

Many different types of coasts are represented along the ca. 7,000 km coast of Sweden. Among the most well-known are the archipelagoes. Gillöga, Stockholm archipelago.

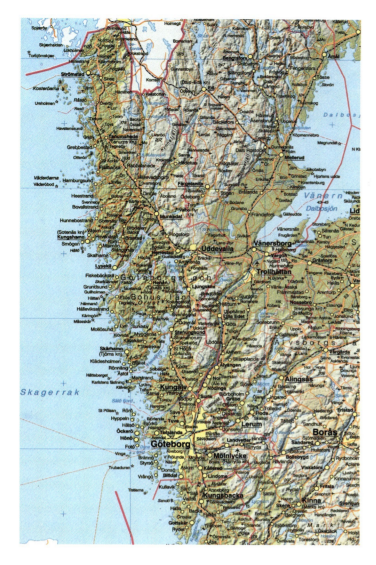

The lower course of the river Anråsälven, to the north of Fjällbacka. The bare rocks, with major structures in the NE-SW direction, are interspersed with fertile valleys of clay soil.

The Archipelago Coast of the Skagerrak

Along its entire length, this coast has a fringe of severely exposed seaward skerries ranging from Onsala in the south up to the Norwegian border in the north. It is only in central parts that the archipelago has a wider depth as a result of the fiords on either side of Orust and Tjörn, and of Gullmarn and Åbyfjorden, which penetrate deep into the hinterland. This is a coastline of undulating bedrock relief which, with its sharp edges and altitudes well above 150 m give a powerful impression. The water depths here are unusual among Swedish archipelagos, perhaps apart from the High Coast. There is frequently deep water all the way in to the mainland, also in narrow sounds, and the greatest depths are exceptional for Swedish conditions: 142 m in Gullmarsfjord and 247 m in Kosterrännan on the landward side of the Koster islands.

Idefjorden, Bullersjöarna and the valley down to Gullmarsfjord cut straight through the landscape in a north-south perspective. To the west of this we find the reddish Bohus granite with dominating SW-NE fissure valleys and others running at right angles to this direction. The fissure tectonics, the vertical jointing and the horizontal banking, the fine fabric of the granite and the sculpturing of the inland ice have created down to the smallest detail a smooth rocky landscape on the scour side of the ice. From Tjörn to the south we find east-west fissure valleys, a pattern which continues in the tectonised granite to the south of Göteborg.

The coast appears to be more exposed to the destructive forces of the sea than it actually is as a result of the extent of rocky land. However, the naked rock is less conditioned by climate than by an original lack of soil. One of the foremost morphological features is also the contrast between the naked rocks and the clay sediment in the fissure valleys. Settlement is another feature of such importance that it forms a true feature of the landscape. In no other place do we find the rocks of an archipelago so dominated by fishing villages.

The gneiss granite of the Koster Islands, with numerous dykes of dolerite, has a pronounced structural relief.

A cliff face with selective weathering of hypoabyssal rock, North Koster (on right).

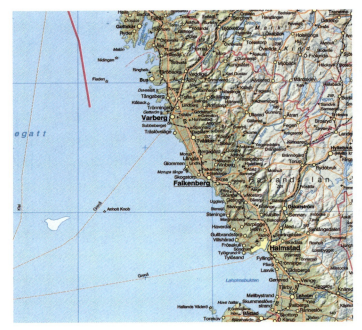

Restored sand dunes at Tylösand. The photo was taken in 1987. The rhythmic pattern of the shoreline has been created by the standing refracted waves.

The Lowland Coast of the Kattegat

The open coast of the Kattegat is also the coast of the Halland plain, an agricultural area which still remains open and with extensive sandy soils, partly planted with protective woodland to prevent drift, a problem which was first solved during the late 19th century. To the north of Halmstad, and from Varberg up to Kungsbackafjorden, the wooded hills nudge out towards the coast.

The coastline of southern Halland consists of attractive curves of sandy beaches. By far the largest is the thirty km beach of Laholm Bay. The entirety of the bay can only be appreciated from the high vantage points of the Bjäre peninsula. The five km long curves of the beaches at Tylösand and Haverdal are almost as impressive when seen from the rocky headlands of Skallen and Tylöudden from the north and south, respectively. On a smaller scale, there are numerous renowned bathing beaches around Falkenberg and northwards up to Apelviken at Varberg. During the 1960's, it was feared that the increased wear to which the dunes lining the sandy beaches were exposed would lead to disastrous erosion by wind and waves. Comprehensive investigations resulted in the initiation of protective measures in the shape of sand barriers, plantation of marram grass and the planning of land use. In this way, the coastal dunes, despite increasing numbers of visitors, could be preserved. In central Halland, mainly from Falkenberg and to the north, we find another type of flat shoreline between sandy bays, shore meadows and boulder-scattered salt marshes. Even flatter seabeds with soft fine sediment and protruding stones are found in the inlets to the north of Väröbacka and sheltered behind Onsala. This is where the low-lying coast comes to an end and the archipelago landscape starts. If the southern coast of the Kattegat is to be regarded as a paradise for motorized sun-worshipping beach-lovers, then perhaps the central parts may be a paradise for bird watchers, and the northern parts a seventh heaven for boat enthusiasts.

The coast at Tylösand in 1972. A wavy system of bars can be seen off the beach. Dune erosion is extensive owing to the wear around Tylösand Hotel. Note the pattern of pathways around the summer cottages.

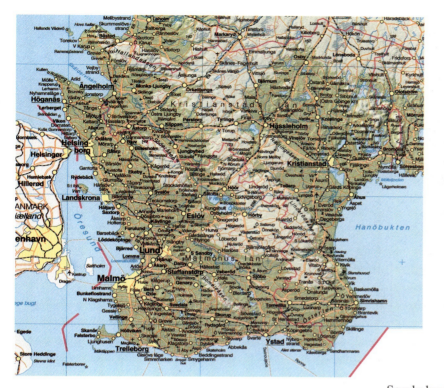

Hovs Hallar is a cliff originating from a basement fault. The highest raised beaches, about 9 m a.s.l., were formed about 4,500 B.C.

The Open Coasts of Skåne

The open coasts of Skåne are extremely rich in contrasts. Shelving sandy beaches in large bays such as the Hanöbukten, Ystadsbukten, Lommabukten, Lundåkrabukten and Skälderviken alternate with massive precipitous coasts of horsts of pre-Cambrian gneiss on Bjärehalvön, Kullen and at Stenshuvud, and with protruding sandy headlands such as Sandhammar and Falsterbohalvön. High cliffed coasts of loose deposits reoccur in the Sound; on Ven, at Helsingborg and Glumslöv, and also along the south coast at Kåseberga. Low moraine coasts run along the south coast. No other region has so many sandy beaches and such active bedrock and till cliffs. No other place has so much shore erosion, and only in this province do we find coasts without ongoing isostatic uplift.

Sandy bays in equilibrium usually have a symmetrical shoreline curvature. Asymmetry is an indication of imbalance in the longitudinal transport of sand. Åhusbukten has a skewed curvature indicating a long-term transport towards the north. Similarly, in Ystadsbukten, there is transport towards the east and erosion in the south and western parts. The material there migrates out of the bays and is lost. If the supply of material from the present erosion area is prevented, then transport in the bays will still continue and the erosion will extend in the direction of the transport. In Skälderviken the net transport is directed to the north where the sand is lost through wave reflection against the breakwater at the mouth of the river Rönneån.

At Sandhammaren, sand is accumulated in balance between the eastward transport past Löderup to the west and the southward transport past Mälarhusen to the east. Large amounts of sand have blown from the shore of Sandhammaren and today form a wide belt of dunes. The lack of material balance to the west has caused extensive shore erosion at Löderup. The erosion protection built there may have a negative effect on the shore closer to Sandhammaren. The general tendency for erosion along the coasts of Skåne may be ascribed to weak land subsidence.

The Falsterbo peninsula has a curious anvil-shaped coastal contour with the striking surface turned to the west. Along this shore the transport alternates between north and south. The latter transport, in conjunction with westward transport along the south side of the peninsula, has created the moving beach berm formation known as Måkläppen. Northward transport along the western shore is effectively hindered by Skanör harbour, which has resulted in the shore to the south of the harbour being widened by more than 400 m since 1860.

Cliff erosion at Rörum, to the south of Stenshuvud.

Comparison of the maps shows how the shoreline has changed as a result of erosion at Löderup.

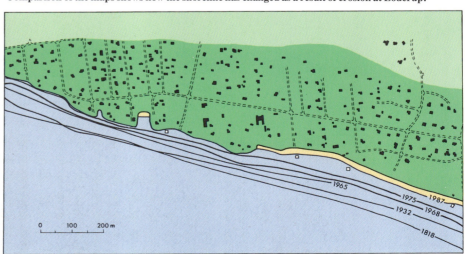

The Blekinge Archipelago

To the east, the Blekinge archipelago is flanked by a southerly extension of the Kalmarsund plain, which at its extremity forms a very flat and shallow, clustered archipelago, which ends at Utlängan and has its outermost extremity at Utklipporna. To the west, Listerlandet and Hanö form an abrupt morphological and geological boundary.

Hasslö, Aspö, Tjurkö and Sturkö are separated by narrow sounds which are further complicated by old naval defence obstructions, within which is Yttre redden, an open bay with good space even for large sailing ships. It was the ideal location for a naval base.

Further to the west, the coastline retreats since the land rises rapidly. Behind the actual coast in Blekinge, there are clear, narrow fissure valleys aligned towards to the NNW and ENE. In the western parts, these valleys form deep inlets which offer opportunities to build the harbours and bases required by a modern navy.

The islands in the Karlskrona archipelago are very low. Heights are rarely above 15 m and in the bays inside them, there are depths of 10–20 m. In the western coastal area, there are two higher offshoots; Tärnö and Stärnö. From recreational and conservation points of view, Blekinge is perhaps mainly known for its luxuriant deciduous woodland which also formerly covered the large seaward skerries but which have been severely decimated as a result of the large population increase.

Ekö island, in the Karlhamn archipelago to the north of Tärnö, in the spring.

Kalmarsund, Öland and Gotland

A pile of Cambro-Silurian sediment (600–400 million years old) lies like a toppled pack of cards which has fallen out from the Kalmar coast to Öland and Gotland and further over to the Baltic States: The further out we get, the higher up in the sequence of strata we come, until down to the southeast the strata bends downwards.

The low Kalmar plain is a continuation of the pre-Cambrian peneplain which along the coast is further levelled-out by overlaid and eroded Cambrian sandstone. Here, we have a dominance of low moraine shores with occasional moraine skerries and islets, which start to develop into an archipelago to the north of Skäggenäs. The foremost characteristic of this coast is stony shore meadows.

In the gently upwards-tilted strata on the west coast of Öland and Gotland, bedrock cliffs have been formed as a result of the inland ice and through subsequent wave erosion during different developmental phases of the Baltic. Examples are the present-day cliffs on Öland, which are composed largely of clay and alun slate along the southern part of the west coast up to Djupvik, but further to the north in orthoceran limestone, which also develops into the majestic western precipice called "Landborgen". The cliffs have been damaged by numerous old quarries. The gently sloping land towards the east runs largely parallel to the limestone strata and creates a flat coastal plain along the southern east coast. To the north of Kapelludden the contours of the coast become interrupted with southeasterly points of moraine. The open curving shoreline of Böda Bay is of a completely different kind.

The gently sloping Silurian sequence towards the southeast is reflected in the large coastal contour features of Gotland and in the major resistant stratum which runs in a NE-SW direction across the island. The longest connected sequence of cliffs is thus found on the northwest coast –

Predikstolen—the Pulpit—at Balsklint at Stenkyrka on the northwest coast of Gotland. Two coral reefs can be seen in the cliff, the one on the right has partly eroded. Sediment has accumulated between the reefs. On the shore, so-called Philip structures have developed as a result of a round lump of reef pressing down underlying layers. The structure became visible when the reef eroded.

Högklint at Visby, 46 m – since the east coast is mainly a low moraine coast alternating with smaller sandy shores. Attractive shingle beaches are found both on the east and the west coasts but the most beautiful of all is the pattern of shingle ridges on Fårö and the small islands.

Marl and limestone do not form sand when eroded. They weather and are ground down into a fine sludge which is carried out into deep water. Sand mainly comes from shield area till and glacio-fluvial sand which is washed out by the waves. An area with large amounts of sand covers the bottom to the north of Gotland, stretching from Kopparstenarna to Gotska Sandön over Salvorev to Fårö and along the east coast. The core of Gotska Sandön consists of till which has been transformed by sea waves and wind. It was first formed as an underwater reef and slowly rose from the sea about 6,000 years ago. The island is dominated by dunes and old shore formations which are covered by pine forest. The shape of the shoreline of this isolated island reflects the balance between different wave directions. At present, erosion is at work along the south and west coasts and there is an extension of the northwestern point.

On eastern Fårö, the sand forms vast fields of dunes which compete with the forest, and here also the finest sandy beaches on Gotland are to be found. To the west, the fields of shingle and sea-stacks dominate.

Kristianopel, at the southern entrance to Kalmarsund. The coastal plain with low till skerries.

Ölands norra udde (Öland's northernmost point) with shore barriers encircling Grankulla inlet. A group of raised beaches can be seen in the foreground.

Recurved spit on the eastern side of Langhammaren, Fårö. The spit has emerged since the late 1920's.

The Archipelago Coast of Östergötland

The coast of Östergötland is formed as a peneplain sloping gently towards the east and south and which is crossed by fissure valleys with sediment bottoms, and between them ridges of rock washed clean of soil. The roughly lobate land contours of this coast are dominated by long and narrow bays and sounds, oriented in a NW-SE direction, such as those at Arkösund, Valdemarsviken, the entire inner archipelago of Tjust, Gunneboviken and others.

The widest archipelago with an extensive outer archipelago is found at St. Anna and Gryt. Alternating fissure directions to the N-S, NNW-SSE and NNE-SSW and an enormous multitude of islets, rocks and skerries protect the large islands of the inner archipelago and offer the yachtsmen innumerable choices in shifting winds. From ground level it is difficult to picture the morphological pattern. The complexity of the bedrock tectonics of this coastal area has only recently been demonstrated through digitally processed models of the terrain.

The surface of the bedrock does not consist of a rigid, level peneplain with cracks suggestive of scratches on a fixed surface. In actual fact, the surface consists of vast blocks covering square kilometres in size, some raised, others lowered and most of them also clearly tipped on one side. These movements have resulted in the fissure zones. In addition, there are smaller fractures within the blocks.

Part of the chart showing the area to the south of Arkösund. The structural direction of the bedrock has created conditions for north-south navigation channels through the islands of the archipelago. A channel marked with lights on the left, and the old "Kronleden" sailing channel is on the right.

The navigation channel in St. Anna archipelago stretches diagonally up towards the left, through dense swarms of skerries.

A satellite view of the Stockholm archipelago, with more than 24,000 islands.

There are few beaches in the Stockholm archipelago. Consequently, Sandhamn ("Sandy Harbour"), with its pilot and customs station, its steamboat connections to Stockholm, and surrounded by open bays, has long been a popular place for bathing.

The Archipelagos of Stockholm and Södermanland

The Stockholm archipelago is usually considered to cover the vast number of islands extending from Björkö-Arholma in the north to Öja-Landsort in the south. Its largest width is more than 80 km, off Stockholm. Here, there are more than 24,000 islands, islets and skerries. From topographical, climatological, botanical, social and economic geographical and other viewpoints, the archipelago is traditionally divided into an east-west zonation in three parts with slightly varying boundaries. The inner archipelago extends from the north across the areas shoreward of Vätö-Rådmansö-Furusund and further through Ljusterö-Lingarö down to Saltsjöbaden, and shoreward of Tyresö-Dalarö-Gålö-Muskö-Torö. The inner archipelago is very similar to the landscape around Lake Mälaren with its thick foliage, oaks, meadows and beds of reeds. From the south, the central archipelago is bordered in the east by the SW-NE barrier of islands through Utö-Ornö-Nämndö-Runmarö-Lökaö, and then towards the NNW up to Blidö and further along the shipping channel past Tjockö and Gisslingö to Arholma. This was formerly the home of the fishermen farmers and the island residences of well-to-do Stockholmers. Everything on the outside of this line is the outer archipelago, with its labyrinthal waters, of which the most beautiful are the "archipelagos of the archipelago" on the shoreward side of Söderarm-Svenska Högarna. Swarms of skerries form necklaces of islands out towards the Åland Sea: Svenska Högarna, Gillöga-Lygna, Fredlarna, Nassa, Björkskär-Röder, Ängskär, Kallskär-Norrpada, Rödlöga, a paradise for thousands of summer yachtsmen.

The boundary lines also reflect the tectonic features in the bedrock. Here, we can distinguish large wavy structures. The entire Södertörn area looks like an enormous question-mark ending with Landsort as the dot in the south. Ingarö and Värmdölandet form another vast curve towards the east. These patterns were created in Archean times by folding and by creation of gneissic granite in the depths of a mountain chain extending from central Sweden to southern Finland and which can also be traced in detail down to a centimetre scale. In contrast, there are the SW-NE linear structures, mentioned above, in the boundary of the island barriers – faults and fissure structures which have created protected navigation channels such as the Möja-Dalarö Channel and the Furusund Channel. Other fissures are at right-angles to these in a NW-SE direction. There are also E-W tectonics such as the fault escarpment along Södermälarstrand in Stockholm. The west side of Södertörn and the Södermanland coast mainly have N-S and NW-SE structural features. Here, the archipelago extends out to meet the open sea.

The Coastal Plain of the Bothnian Sea

To the south of the High Coast, the boundary of the Norrland terrain leaves the coastline and is replaced by an undulating sub-Cambrian peneplain, becoming increasingly flat and wide to the south. In the southernmost third of this area the relief is less than 20 m, in the central third 20–50 m and in the northernmost third 50–100 m. The 100 m high Hornslandet, with water depths of 50–60 m nearby, diverges from the general pattern.

The abundance of large lakes to the south of Söderhamn, close to the coast, is an interesting characteristic. About 7–8,000 years ago, when the land was 50–60 m lower than today, there was an extensive archipelago here, many kilometres wide, and with vast areas of protected waters. Along its northeastern boundary, there is an area of massive NW-SE fissure tectonics, whereas the islands and headlands to the south are completely dominated by a bedrock structure running in an east-west direction.

Along the coast of Gästrikland, which in other respects is extremely irregular, there are a number of smaller archipelagos, dominated by an abundance of islands and skerries. A special characteristic of the shores are the large boulders and boulder-rich moraines. Shores with stones and boulders dominate.

The flat, horizontal peneplain of Uppland forms the southern coast of the Bothnian Sea. To the east of the mouth of the river Dalälven, Billudden, the northernmost land part of the Uppsala esker, protrudes into the sea, being continuously re-designed by land elevation and waves. In the shallow Lövsta Bay, low moraine hummocks form swarms of islands. The easternmost coast of Öregrundsgrepen has a low, pointed contour, severely exposed to the northerly winter storms. Gräsö-Singö-Väddö-Björkö-Arholma and the intermediate smaller islands and skerries form a NNW-SSE barrier between the southern Quark and inner Roslagen, where a fissure system oriented in the same direction is crossed by east-west open valleys, formerly important navigation routes to the central parts of Uppland.

Billudden is the northernmost part of the Uppsala esker in the Bothnian Sea, close to the mouth of the River Dalälven. The old fishing village of Nothamn was built in a sheltered inlet, between the crown of the esker to the west and the wave-created recurved spit to the east.

Enskär fishing village, to the south of Skatön. The coast in Gästrikland is very rich in large boulders and the land elevation of 60 cm per century leads to rapid changes in navigation.

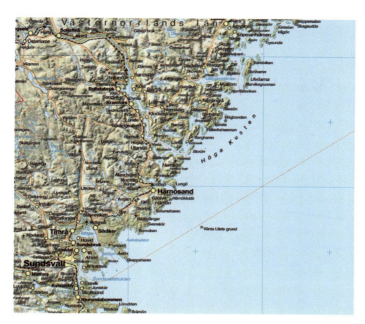

The High Coast – Höga kusten

The strongly undulating hilly terrain of the hinterland, with height differences of more than 100 m, breaks through to the coastline at Skagsudde. Along the coastline known as Höga Kusten – the High Coast – which extends down to the mouth of the river Ljungan, the relief reaches heights above 300 m. The mean depths in the sounds and the bays is 30 m and the maximum depth 154 m. The undulating hills form a rectangular pattern mainly as a result of a system of fissure valleys running north-south and east-west. This system is abruptly terminated by the NE-SW coastline. The central areas of this coastline are Skuleskogen and Nordingrå. Steep rocky faces without beaches make large parts inaccessible. The position of the highest shoreline, formed during the melting of the inland ice, is found at Skuleberget, 285 m above the sea. The great water depth at the time when the highest coastline was formed, and during subsequent land uplift, led to hill summits becoming exposed to large waves which washed off the till and created vast shingle beaches with additions of frost-weathered boulders.

Three of Norrland's mightiest rivers open into this area: Ångermanälven, Indalsälven and Ljungan. The delta front of the river Ångermanälven is at Nyland, more than forty kilometres from the mouth of the inlet at Hemsön. The delta is almost completely water-covered and the inlet has a maximum depth of 80 m.

Sweden's only well-developed, large coastal delta is in the mouth of the river Indalsälven at Klingerfjärden. The river makes a curious turn towards the north before reaching the delta. The pre-glacial river had its outflow straight to the southeast over Timrå. The old river valley, with its bedrock bottoms at 50 m depth beneath the present-day sea level, has been filled up with glacial deposits. At the present rate of land elevation, the surface of the sea will come down to that level in 7,000 years time. The development of the delta can be followed on maps since the 17th century. The catastrophic, man-made emptying of Lake Ragundasjön in 1796 carried large amounts of sediment down to the delta and the main water flow shifted from the main channel in the east to that in the west. Since the early 19th century, the front of the delta has advanced by about 1 km. Today, the delta is supplied with negligible amounts of sediment from the river stretch upstream of Bergeforsen, owing to the hydroelectric dam and river regulation. The river channels are partly covered with erosion protection and linings from the time when floated timber was sorted in the central channels.

Bare rock summits washed clean of soil all the way up to the highest shoreline at 285 m a.s.l. can be found at Skuleskogen in the northern part of Höga Kusten.

The delta of the River Indalsälven at Klingerfjärden. The estuary bank is partly bordered by broken pallisades. A network of partly deserted channels can be seen in the background.

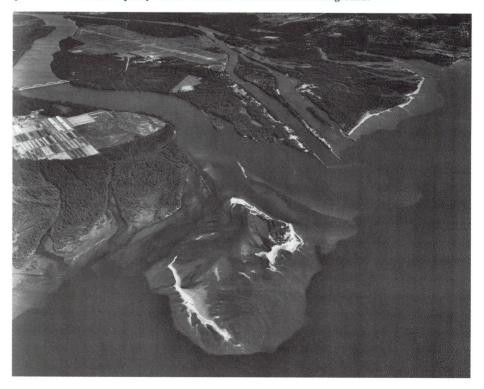

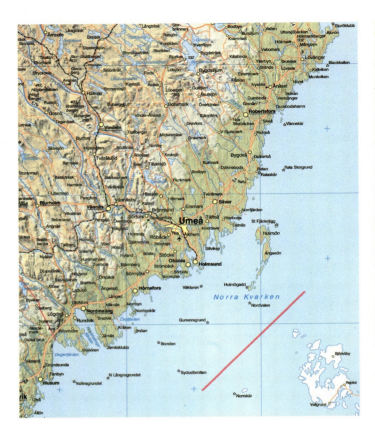

The Coast of the Northern Quark

To the south of Skellefteå there is an abrupt shift in the coastline towards the southeast over about 50 km. The coastline from the mouth of the river Pite älv past Skellefteå and down to Bjuröklubb may be regarded as a transitory zone from the wide archipelago coast of Norrbotten to the open low coast along the northern Quark. Here the seabed is not bent downwards but forms a shallow strip of land projecting towards Finland with a sill at about a depth of 30 m.

The northern part of the coast towards the Quark is characterized by a low hilly terrain with a relief of 20–50 m. A series of faults have tipped the old plane bedrock surface at an angle and cut off the drainage from the inner parts of the country. The shoreline forms a system of open inlets with widths from 100 m to 1 km. Along the seventy km long coastline from Bjuröklubb down to the south of Ratan there are about seventy inlets, many of them with a small rivermouth. The rivers follow the linearity of the bedrock structure in a NW-SE direction, which further to the south increasingly turns to a north-south direction in the extremely flat Västerbothnian coastal plain. This is reflected in the parallel, narrow lobate form of the shoreline and in the elongated islands, with additions of till forming drumlins.

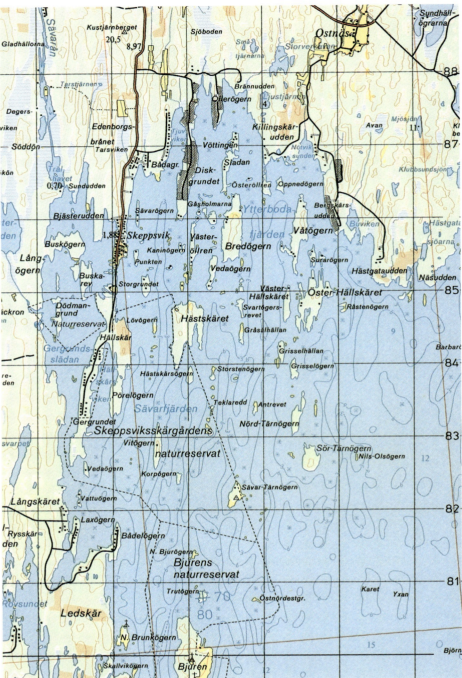

The coastal plain to the south of Sävar. A pronounced north-south bedrock structure with superficial deposits of till gives the coast its characteristic appearance. Map of Umeå, 20K NO, scale 1:50 000.

View from Täftefjärden to the east over the extremely flat and low coastal plain towards the Quark.

A field of raised shingle beaches on Sandön island off Luleå at 30 m a.s.l.

The Archipelago Coast of the Bothnian Bay

Sweden's northernmost archipelago coastline is characterized by large, low islands among the seaward skerries, large areas of inner protected waters and deep penetration of major river valleys. The coastal plain to the east of Kalix has very low relief. The large-scale morphology further west has more the character of a submerged, well-developed riverine landscape which, in its outer parts, has been bent down and submerged by the sea. That this indeed has happened is confirmed by, for example, the pre-glacial channel of the river Lule älv, which can be followed along the seabed over to Finland. The orientation of the river valleys is reflected in the main shoreline pattern. To the south of the river Pite älv the width of the archipelago decreases considerably as a result of the higher level of the inland. From Kåge, the coast has a dramatic change in direction towards the southeast. Here the coast runs parallel with the structure of the landscape, which creates elongated shoreline patterns.

In large parts of the archipelago and the neighbouring mainland the relief is very low (<20 m), which leads to a rapid growth of land through uplift and subsequent forestation. A characteristic of this area is that all original towns have had to move and new harbours have had to be built closer to the sea on reclaimed land in the mouths of the river valleys.

Glacio-fluvial deposits in and adjacent to the large valleys have created extensive areas of sand. The soil layers have been wave-washed and have formed new beach deposits and coastal forms during the land elevation. By comparisons of maps, it is possible to follow the coastal progression over recent centuries.

The Sandgrönnorna form the outermost, emerging part of the Luleå esker. The development of the shoreline from 1790 until 1985.

A view over the outermost part of the Sandgrönnorna at low water. The effect of wind can be seen on the shoreward side of the beach zone.

Weather and Climate of the Sea

A ribbon of sunlight over the sea. Utö, in the Stockholm archipelago.

The weather brings life to the sea, both figuratively and in a purely literal sense. The complicated interaction between sea and atmosphere is an important component in the sensitive balance which is the condition of life on Earth.

Wind

There can be no doubt about which meteorological parameter most dramatically affects the sea – the wind. The difference between the glittering surface of the sea during the summer and the boiling cauldron that meets us during a full snow storm is immense.

The wind is created by unevennesses in the distribution of air pressure. The air moves into areas with low pressure from areas with high pressure; the greater the difference in pressure, the faster the movement. However, these movements do not take place directly towards the lower pressure area since the rotation of the Earth around its own axis deflects the movement of the air so that instead it becomes almost parallel with the lines for similar air pressure – the isobars. In the northern hemisphere, the air moves so that the lower pressure is to the left of the wind direction. In the lowest part of the atmosphere, however, friction with the Earth's surface always causes a certain flow of air towards areas with lower pressure. Since friction is greater over land than over sea, the angle between the wind and isobars will also be greater over land than over sea.

The greater friction over land also causes the wind speed there to be lower than over the sea. At the coast itself, the wind speed and the wind direction may vary in an extremely complicated manner. The topography of the coast often steers the wind in the direction of the fiords and sounds at the same time as the wind may be much stronger around headlands and where the sounds become narrower.

The distribution of air pressure is much more uniform during summer than during winter in the waters around Sweden. In January, the average difference in pressure between the southernmost Baltic and the northernmost Bothnian Bay amounts to about 4 hectopascal, with the lowest values in the north. In July, the corresponding difference is not even half as large. Over the sea, it is also windier during winter than during summer, when the frequency of gales and storms is much lower than during winter.

In agreement with the average air pressure distribution, winds between south and west prevail in our part of the world, although wind direction varies considerably in a climate such as ours. The wind changes direction in pace with the passing of the low pressure systems. An interesting feature of the distribution of wind direction is shown by the wind roses from stations along the west coast. There is a certain dominance for winds between north and east during the winter. This maximum is explained by cold heavy air from the inner parts of Götaland being drained out over the coast. The corresponding effect is also found along the east coast where, however, this local flow of air coincides with and is hidden by the dominating southerly to westerly winds.

Another well known local wind system is the land and sea breezes along the coasts on spring and summer days, when temperature differences between the surface of the sea and the sun-heated soil surface are greatest. Sea breezes may be strong, with wind velocities of 5–10 m/s in some cases. The breeze moves further in-

A sea breeze occurs when sun-heated air over the land rises and is replaced by cooler air from the sea.

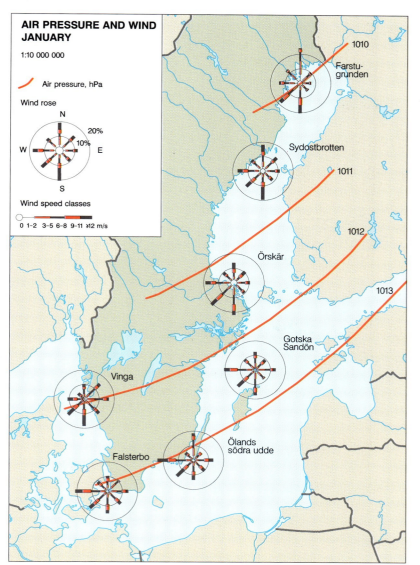

The distribution of air pressure is analysed by means of material obtained between 1931 and 1960. The wind roses show the monthly mean values for 1961–1989. (G10, G11)

The autumn is a windy period. On the West Coast, gales occur on at least one day out of four during November. The diagrammes are based on material from 1961–1989.

WIND AT SEA

Strength Beaufort	Velocity m/s	Description	State of the sea surface
0	0–0.2	Calm	Mirror-like sea
1	0.3–1.5	Light air	Small-scale ripples
2	1.6–3.3	Light breeze	Small wavelets, still short but more pronounced, crests do not break
3	3.4–5.4	Gentle breeze	Crests begin to break, foam of glassy appearance
4	5.5–7.9	Moderate breeze	Waves still small but becoming longer, fairly frequent white foam crests
5	8.0–10.7	Fresh breeze	Moderate waves taking a pronounced long form, many white foam crests
6	10.8–13.8	Strong breeze	Large waves begin to form, crests break and form large areas of white foam
7	13.9–17.1	Near gale	Sea heaps up, white foam begins to be blown in streaks in the direction of the wind
8	17.2–20.7	Gale	High waves with crests of considerable length, foam is blown in dense and well-marked streaks in wind direction
9	20.8–24.4	Strong gale	
10	24.5–28.4	Storm	Very high waves with long overhanging crests; foam causes sea surface to appear white
11	28.5–32.6		Exceptionally high waves, sea surface completely covered with white foam which also fills the air and reduces the visibility
12	32.7 and above	Hurricane	

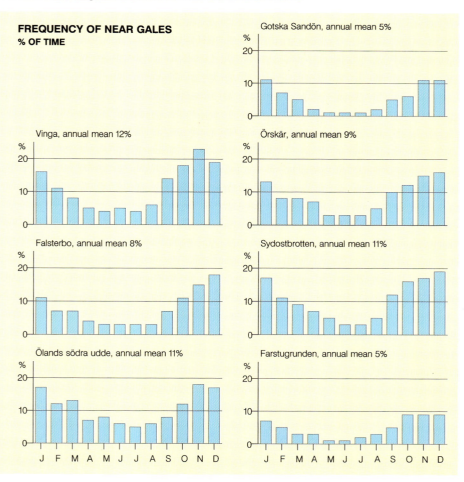

land as the day gets older. At the same time, the direction of the sea breeze turns clock-wise, whereby the direction at right-angles from the sea and in towards the coast in the morning becomes increasingly parallel with the coast in the evening. There may be a dramatic reduction in temperature when the sea breeze arrives. The land breeze, from land out towards the sea, is usually not as strong as the sea breeze.

A combination of low temperature and hard wind is, in many respects, a dangerous event. In such weather, people may suffer frost-bite even if the temperature has fallen only a few degrees below freezing point. As long as the sea is open, the combination of cold and strong wind also implies the risk of severe icing on vessels and fixed constructions at sea. Spray from the waves freezes into ice and the process is so rapid that hundreds of tonnes of ice can be formed in a few hours.

Sunshine and Clouds

Because the radiation from the sun and from terrestrial sources (atmosphere, soil surface, etc.) originate from two well-separated ranges of wave-lengths, distinctions are made in meteorology between short-wave radiation (0.29–3 micrometer range) and long-wave radiation (3–100 micrometers).

Short-wave radiation on a horizontal surface is called global radiation. This can be divided into direct and diffuse solar radiation. Some of the short-wave radiation is reflected and this part is proportional to the albedo (the reflectance of the surface).

Long-wave radiation is extremely diffuse and depends upon, e.g., temperature and cloudiness. Distinctions are drawn between the downward and the upward component. The latter can be calculated using the Stefan-Bolzmann law if surface emissivity and temperature are known. The radiation balance consists of the difference between incoming and outgoing radiation, whereby both short- and long-wave radiation are taken into consideration.

When visiting the coast during summer, many of us have certainly often noticed that it is much sunnier out to sea than over the land. In the morning, there may be fine weather over land but as the sun starts to get hotter, clouds start to appear, whereas the sun is still shining out to sea. This is because the air closest to the sun-warmed surface of the land also warms up, and thus becomes lighter and rises upwards before again becoming cooled. During cooling the water vapour in the air is condensed and forms clouds.

Fog may also occur when the air is cooled so that condensation occurs. The only difference between clouds and fog is that the latter is defined to extend down to the soil surface. The limit between fog and haze has been placed rather arbitrarily at a visibility of 1,000 m. The most important type of fog over the sea occurs when warm air is cooled by a cold water surface. This often occurs during spring and early summer, for example when warm air from Russia blows out over the Baltic which is still cold. A similar type of fog may also occur over upwelling areas with extremely cold water close to the coast when there is a wind blowing from land. In addition, in situations of severe cold and light winds, there may occur sea smoke over sea areas which are still unfrozen. Since temperature differences between sea and land are largest during spring, the frequency of fog at the coast and over the sea is largest at that time, in contrast to the situation on land, where fog more frequently occurs during autumn. Another important difference between sea fog and fog on land is that the sea fog does not have any clear variation throughout the day, whereas fog on land is most common during the night and early morning.

Icing may have disastrous effects even on large ships. The risk is greatest when there is a combination of low temperatures and strong winds.

Sea smoke

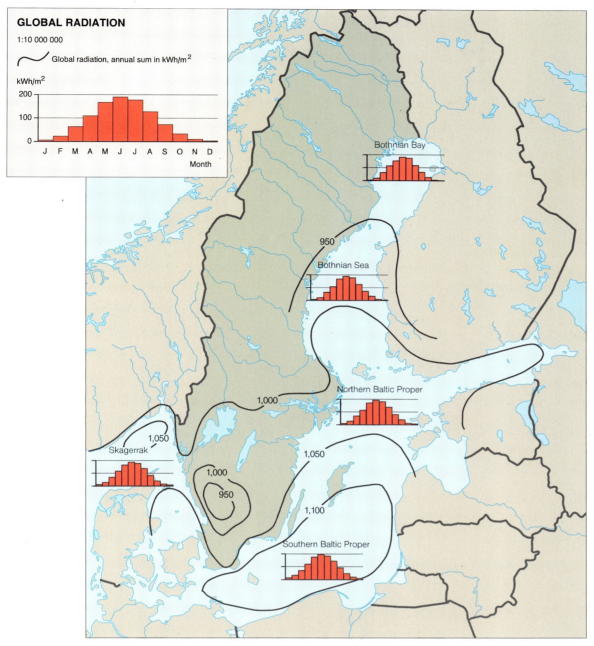

The monthly mean values are based on data from the period 1961–1983. When calculating the annual sums consideration is taken to the average ice and cloud conditions and to measured global radiation. (G12)

Air temperature over the sea, monthly mean values for the period 1931–1960. (G13, G14)

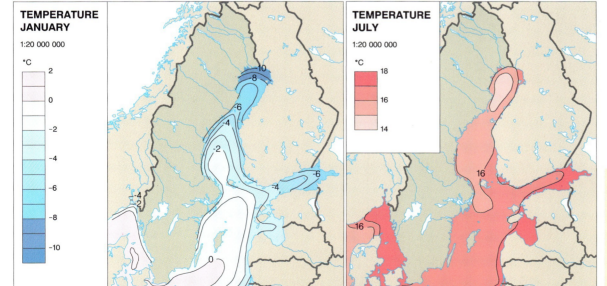

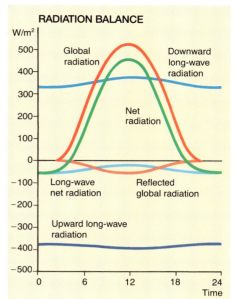

Daily variation in radiation balance above the Baltic Sea in July.

Temperature

The temperature climate over the sea is characterized by small variations both during the year and during the day. The sea thus has a strongly moderating influence on climate. Temperature conditions in Sweden are, in addition, characterized by our position on the west side of a continent and by the dominating westerly winds, which give us a very favourable temperature climate in relation to the northerly location. This particularly applies during winter when air temperatures over marine areas around Sweden are about 20°C higher than over Hudson Bay in Canada, to take just one example from another marine area at about the same latitude as ours.

In January the mean temperature of the air is about zero over the Skagerrak, Kattegat and southern Baltic, where there is almost always open water throughout the winter. The mean temperature decreases towards the north but is still as high as −2°C above the central part of the southern Bothnian Sea, which is also a result of ice-cover being relatively unusual there in January. Over the Bothnian

Difference in daily mean temperatures between Söderarm and Norrtälje.

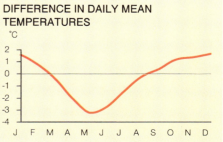

The meteorological station at Svenska Högarna. More than 200 people throughout Sweden make the daily weather observations used in the production of forecasts for radio and television.

Sea and Bothnian Bay the air temperature rapidly decreases and is −5 to −6°C over the Northern Quark and ca. −10°C in the northernmost part of the Bothian Bay.

The variations are considerable from year to year depending on the ice situation. We must remember that at least the Bothnian Bay is completely covered by ice and snow during severe winters and then may be regarded as a continuation of the vast Eurasian continent to the east. In such situations the cold winds from the northeast and east reach Sweden without being warmed during their passage over the sea.

In spring, when the sun and warm winds cause high temperatures over land, it is still cold out to sea, and this difference in temperature remains until August, by which time the temperature of the sea air has caught up.

During summer, there is only a small difference in mean temperature between land and sea, and even geographical variations are negligible. The warmest parts are over the inner Skagerrak and over the Kattegat and the Sound, where the mean temperature is about 17°C in July. The temperature along the Baltic coasts is also 17°C at that time, whereas it is about 16°C along the coasts of the Bothnian Bay and slightly lower than 15°C in the central parts of the Bothian Sea.

The enormous quantities of energy stored in the sea during the summer are released only slowly during the late summer, autumn and early winter, and thus it is still relatively warm at that time out to sea. The rate at which heat is released from the water depends on the wind velocity and the temperature difference between the surface of the sea and air. The fastest rate of release is when the air temperature is low and the wind is strong.

Apart from direct transport of heat, there is a very important exchange of latent heat between the sea-surface and the air above. This energy exchange takes place by water evaporating from the surface of the sea and then condensing into fog or cloud droplets. During condensation there is a release of large amounts of heat to the atmosphere.

Precipitation

Precipitation can be divided into two main types, showers and more general precipitation. In the part of the world where Sweden is located, the latter type occurs in conjunction with the fronts of low pressure, between cold and warm masses of air.

In general, there is less rain and snow over sea than over land, but the reverse situation is also found. One such case is when extremely cold air moves out over open water. The air is then heated in the lower layers which causes it to become extremely unstably stratified. The warm air is light and rises, whereby it is cooled, with resulting condensation and release of precipitation. In such situations, there will frequently be heavy snow showers over the open sea and neighbouring areas of land.

During summer, showers frequently occur over land when the surface of the ground is heated by the sun during the day. Consequently, during summer months, there is considerably better weather at sea than over land during the afternoons. The fact that more precipitation generally falls over land also during the rest of the year depends on air being more turbulent in such places than over the sea, which in turn depends on the surface of the land being considerably more uneven than the sea-surface. Particularly in very hilly terrain there will, thus, be considerably larger amounts of precipitation than over the sea.

The distribution of precipitation over the sea is known only in relatively approximate figures owing to the sparse network of stations and to the difficulty of measuring precipitation in places exposed to the wind. In general, precipitation is greater in southern than in northern waters around Sweden, and in addition it is frequently greater closer to the coast than further out to sea.

Marine forecast areas used in the SMHI reports broadcast by Swedish Radio. (G15)

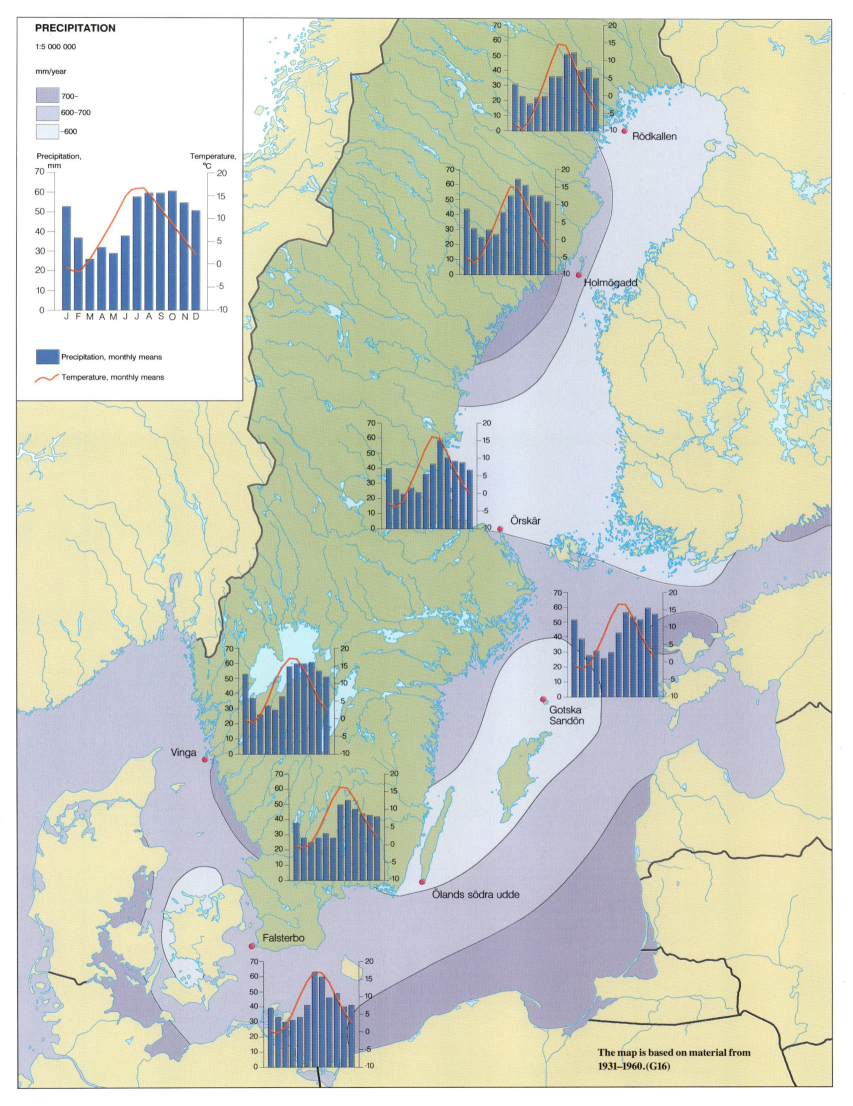

The Sea and Sea-water

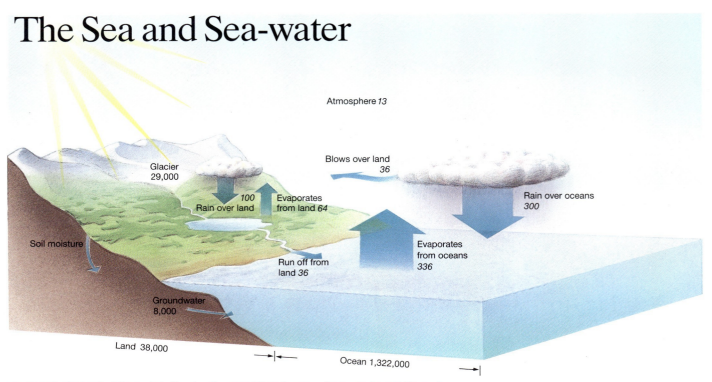

The hydrological cycle. Figures in italics give the annual flows between the reservoirs and figures in Roman type give the size of the reservoirs. The unit is 10^{15} kg. Of water on land, as much as 77% is frozen (glaciers and polar ice).

Water probably has its origin in the depths of the Earth, from which it still runs into the sea. Water is in continuous circulation between sea, atmosphere and land. On the basis of the relationship between sea volume and the amount of water entering the sea each year in the form of precipitation and river water, the average retention time of water in the sea can be estimated at 4,000 years. The corresponding figure for the atmosphere is about 10 days. The sea-water contains dissolved salts. Salinity is given with the unit PSU meaning practical salinity unit, it is defined so that numerically it approximately gives the amount of salt in water in parts per thousand by weight. Salinity levels in the various oceans are relatively similar and mean salinity is about 35 PSU, which means that of 1,000 kg sea-water, on average 35 kg consists of dissolved salts. The elements which together make up the salt are largely present in the form of simple and complex ions. As regards weight, sea salt is dominated by a small number of elements, the main components. In addition to these, there are practically all known basic elements present in very small concentrations (trace elements).

Each element in the sea goes through its own circulation. They are supplied to the sea from the land, from the atmosphere and from the seabed and are removed by biological uptake, sedimentation and precipitation. The turnover times in these circulations vary from years to millions of years. Elements with a long turnover time in comparison with the water itself are present in mutually constant relationships regardless of the total amount (the salinity of the water). This applies, for example, to the main components. Some of the trace elements, so-called nutrient salts, are very active biologically. Among these we find different nitrogen compounds, phosphate and silicate. Concentrations of these elements vary within wide limits, both from place to place and, in the surface layer of the water, from time to time. Sea-water also contains dissolved gases. Most of these gases enter the sea from the atmosphere. Some gases are strongly associated with biological circulation. This mainly applies to oxygen, carbon dioxide and hydrogen sulphide.

Sea-water density depends primarily on salinity and temperature. Fresh water has its highest density, 1,000 kg/m^3, at +4°C. Colder and warmer water has a slightly lower density. Sea-water has a higher density than fresh water on account of its salinity. In the open sea, the density is about 1 027 kg/m^3. The temperature at which salt water has its highest density is lower than +4°C and decreases with increasing salinity. The density of the water also depends on the pressure, i.e., how far below the surface of the sea the water is located.

The geographical differences in density of sea-water which result from differences in temperature and salinity are of decisive importance for movements in the seas of the world.

Distribution of Different Water Masses

Sea-water gets its characteristics in a thin surface layer whereupon water masses with different salinities and temperatures are formed. The salinity is determined on the basis of evaporation/precipitation and supply from rivers in combination with the vertical mixing in the surface of the sea caused by the wind. The surface water is relatively low in salinity at the equator where precipitation is high. Salinity increases towards the sub-tropical high pressure areas where evaporation is large, and then decreases towards high latitudes. Divergences from this simple dependency on latitude are considerable and may be explained by the large-scale system

SOME ELEMENTS IN SEAWATER

Element	Kg/tonne seawater (salinity 35 PSU)	Total amount in millions of tonnes
Carbon	0.28	37,000,000
Oxygen	0.006	7,930,000
Phosphorus	0.00006	79,300
Mercury	0.00000003	40
Gold	0.000000004	5

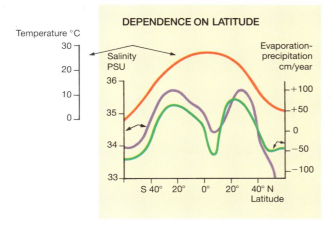

The curves represent a mean value for all oceans.

THE MAJOR IONS IN SEAWATER

Element	Chemical symbol	Kg/tonne seawater
Chloride	Cl^-	19.3
Sodium	Na^+	10.8
Sulfate	SO_4^{2-}	2.7
Magnesium	Mg^{2+}	1.3
Calcium	Ca^{2+}	0.4
Potassium	K^+	0.4
Bicarbonate	HCO_3^-	0.1
Bromide	Br^-	0.1

Kg/tonne seawater with a salinity of 35 PSU.

of currents in the oceans, atmospheric circulation pattern and additions from rivers.

The temperature of the surface water is determined through supply of heat through the surface of the sea and through transports in the sea. Heat is mainly supplied from direct solar radiation, which is absorbed in the surface layer of the sea. As an average over the entire world and over the whole year, there is a balance between in-coming and out-going heat. On the other hand, there is an imbalance between low and high latitudes, between summer and winter, and between day and night. The temperature of surface water is high at the equator and decreases towards the high latitudes as a result of the global imbalance in insolation. In addition, surface water temperature reflects how the large current systems distribute water with different temperatures. At low latitudes, heated surface water flows towards high latitudes where it is cooled. Most of this water returns to low latitudes as cold surface currents. a small amount sinks as a result of the cooling and forms the deep water of the oceans. The largest amount of the deep water is formed within a small number of areas of limited size where low temperature combined with relatively high salinity gives particularly high density. One such important area is the sea to the east of Greenland to which the currents take water with high salinity. The total volume of deep water formed annually is such that the water in the oceans of the world has a turn-over time of 500 years.

Beneath the warm surface in low and middle latitudes, the temperature decreases rapidly with depth. The area where the temperature decreases is usually called the permanent temperature thermocline. Low temperatures in deep water show that this water originates from the cold surface layer at high latitudes.

At high and middle latitudes, there is a large difference in heat flow through the sea surface between summer and winter. This difference leads to seasonal variations in temperature within a surface layer down to a depth of 100–200 m. Close to some coasts there are occasionally abnormally low surface temperatures depending on upwelling of colder water from deeper layers which is caused by the wind.

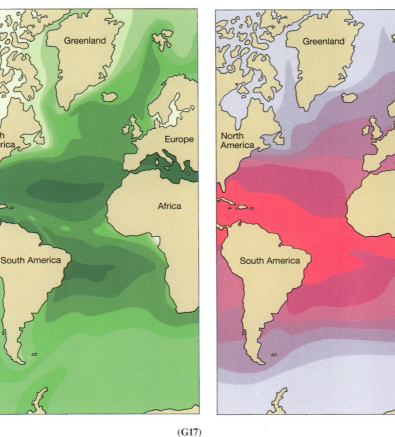

(G17)　(G18)

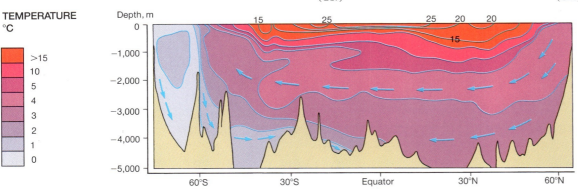

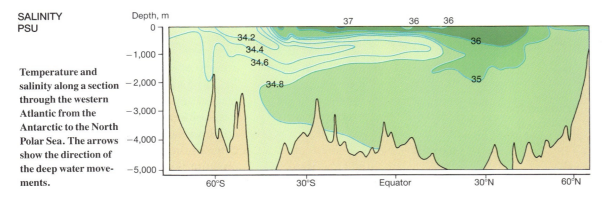

Temperature and salinity along a section through the western Atlantic from the Antarctic to the North Polar Sea. The arrows show the direction of the deep water movements.

The ice-breaker Ymer in the Arctic ice during the Ymer–80 research expedition.

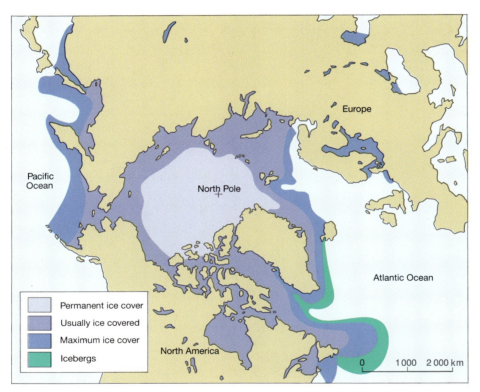

The extent of the ice in northern waters.

Large cavities are often found on the underside of sea ice.

Ice in the Sea

Sea ice consists of crystals of pure water and a varying amount of saline brines enclosed in cavities between the crystals. When sea ice is formed, most of the salt remains in the water and this gives it an increased salinity. The amount of salt enclosed in the ice decreases with time, the saline solution "melts its way" through the ice. Old sea ice thus contains very little salt.

An essential condition for the formation of ice is that the cooled surface water is lighter than the underlying water and therefore remains on the surface for further cooling down to freezing-point. In fresh water this occurs by water with a temperature below +4°C becoming lighter when cooled. In the sea, ice formation can only occur if the surface water has a lower salinity than the underlying water and therefore remains on the surface. This basic condition is satisfied in most coastal areas in temperate zones where fresh water enters the sea from rivers. We have earlier mentioned that the underlying sea-water becomes more saline when ice is formed. This implies that the surface layer with less salt water present initially may, after a certain period of ice formation, disappear so that the conditions for continued expansion of ice are no longer present. If this occurs, the ice may instead start to melt since warmer water comes up from below. Under such conditions, large areas of open water may be formed during the winter.

In most places, the ice disappears during the summer as a result of melting. However, in large areas of the North Polar Sea, the growth of ice during the winter is much greater than the melting during the summer. The net growth is exported by means of ice being transported out of the area, mainly with the East Greenland current.

Currents and Circulation

There are two basic types of currents in the sea; geostrophic current and Ekman current. These are intimately associated with the three horizontal forces influencing water: horizontal pressure gradient, friction and Coriolis force. The horizontal pressure gradient has effect from high towards low pressure. It consists of two parts, a barotrophic and a baroclinic part.

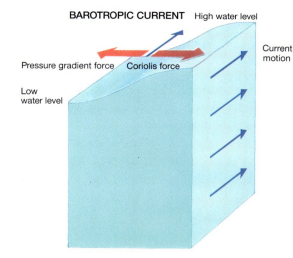

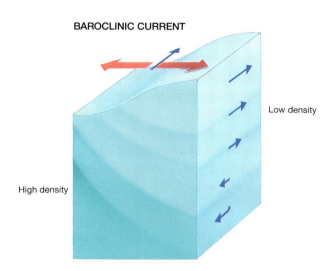

Geostrophic current. If you stand with the surface current flowing against your back, you will have a higher water level on your right-hand side. When the water is stratified the current changes with depth.

The barotrophic part largely depends on differences in sea surface height and is independent of the level below the surface of the sea in question. The baroclinic part depends on the horizontal differences in density and is different at different levels. The pressure gradient is dominant in almost all types of currents. Friction has a levelling-out effect on velocity differences. It transfers the force of the wind into the water and distributes it through a depth of 10–100 m. Beneath this thin layer, in comparison with the depth of the oceans, the velocity differences are generally so small that friction is of no importance in relation to the other forces. The layer where friction is important is usually called the Ekman layer after a Swedish physicist and oceanographer, Vagn Valfrid Ekman. The Coriolis force, named after a French mathematician, is a dominating force in large-scale currents. It is an effect of the Earth's rotation, acting perpendicular to the right of the movement (to the left in the southern hemisphere) and has a strength which is proportional to the velocity and is also dependent on latitude. Large-scale circulation in the sea takes place in conditions of almost complete balance between the horizontal pressure gradient and the Coriolis force. The current is said to be geostrophic. The water then flows parallel to the isobars, with high pressure to the right and low pressure to the left (the reverse in the southern hemisphere).

In situations of wind-generated currents, Ekman currents, the Coriolis force is balanced by the friction caused by the wind. The wind has an effect on a surface layer, the Ekman layer, the thickness of which varies with wind strength and latitude. The theory on which Ekman currents are based, states that the current in the surface is directed 45° to the right of the wind direction (to the left in the southern hemisphere). The strength of the current decreases with depth at the same time as it rotates clock-wise. This pattern of currents implies a transport of water, the Ekman transport which, when summed over the entire Ekman layer, is directed 90° to the right of the wind direction and is an important link to the deeper wind-driven currents.

We have earlier discussed how the geostrophic current is linked to a horizontal pressure gradient, i.e., with a horizontal difference in sea water height and/or a horizontal difference in density. The geostrophic current is generally activated by a small pressure gradient after which both the pressure gradient and the Coriolis force grow in size through interaction which continues until equilibrium is achieved. This phenomenon is usually called spin-up. The original pressure gradient may occur for different reasons: horizontal differences in evaporation and precipitation, or supply from rivers; horizontal differences in warming and cooling; or as a result of horizontal differences in Ekman transports. Horizontal differences in Ekman transports occur out in the open sea if the wind is different at different places. A wind that blows along a coast causes a corresponding effect. As a result of this spin-up, currents at all depths in the oceans are accelerated. Thus, wind which varies from day to day causes currents at much greater depths than the approximately hundred metres which can be reached by wind friction.

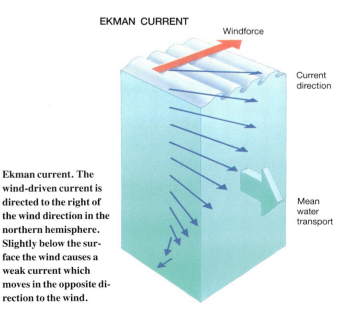

Ekman current. The wind-driven current is directed to the right of the wind direction in the northern hemisphere. Slightly below the surface the wind causes a weak current which moves in the opposite direction to the wind.

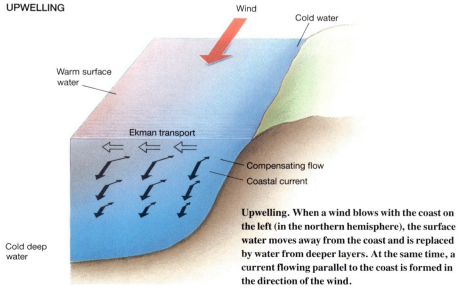

Upwelling. When a wind blows with the coast on the left (in the northern hemisphere), the surface water moves away from the coast and is replaced by water from deeper layers. At the same time, a current flowing parallel to the coast is formed in the direction of the wind.

SURFACE CURRENTS IN THE ATLANTIC OCEAN

→ >0,75 m/s
→ 0,25–0,75 m/s
→ <0,25 m/s

Temperature °C
- >5
- 5–1
- 1–−1
- −1–−5
- <−5

1 East Greenland Current
2 Norwegian Current
3 Labrador Current
4 North Atlantic Drift
5 Gulf Stream
6 Guiana Current
7 Equatorial Current
8 Brazil Current
9 Benguela Current
10 West Wind Drift
11 Barents sea
12 Norwegian sea
13 North sea

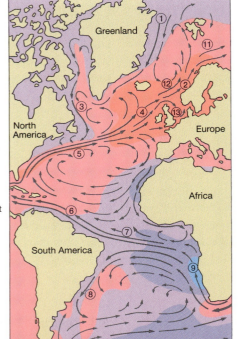

CURRENT SYSTEM OF THE ATLANTIC

The circulation of surface water in the Atlantic can be illustrated in the following way: We have an equatorial system of currents centered around the climatological equator, which is positioned slightly to the north of the geographic equator. In the northern and southern trade wind areas the water flows largely from the east towards the west. In the calms of the equatorial region, there is a weakly developed current in the opposite direction. The water moving towards the poles twists in an anticyclonic direction around the sub-tropical high pressure areas. These vortices are unsymmetrical in an east-west direction. The currents on the west side moving

Surface currents are of great importance for the climate. The temperature lines show how the currents distribute the water with different temperatures. Positive values indicate an increase in temperature in relation to latitude.(G19)

towards the poles are concentrated and have high velocities whereas the return towards the equator is spread out over the ocean. In the belts of westerly winds the currents go from west to east. On the polar side of the westerlies, the water moves in cyclonic vortices with the westerly current at the highest latitudes where the winds are mainly from the east. In the southern Atlantic, the cyclonic vortex is weakly developed since there is open water around the Antarctic so that most of the water in the westerly drift can pass round the globe. In the North Atlantic, the cyclonic vortex is split up into several parts on account of the land masses in that area. The cyclonic vortex in the North Atlantic is of particular climatological interest to Sweden. It carries warm water up to the Norwegian Sea and thus contributes to the mild climate we have in relation to our latitude. The North Sea and the Skagerrak also benefit from this situation as warm water enters at Shetland and passes through Swedish waters. In certain parts, the other oceans have corresponding systems to those in the Atlantic.

The large system of currents is driven by the wind in conjunction with heating and cooling, and with evaporation and precipitation. We do not know how much these different components imply within different areas. As regards wind, there are well-developed theories with which the major features of the oceanic circulation can be introduced. Thus, for example, we can explain the east-west asymmetry shown by the Gulf Stream, and details in the equatorial current system, by the effect of the wind in combination with the curvature of the Earth's surface. Clear evidence of the important influence of wind is found in the northern part of the Indian Ocean. Here, the wind is dominated by monsoon effects with changes in wind direction twice a year. The sea responds by changing its circulation direction within the space of a few weeks.

CURRENTS IN THE NORTH SEA

The water in the North Sea moves in a cyclonic direction. It enters mainly between Scotland and Norway. Most of it flows to the south along the west side of the Norwegian Channel and passes the Skagerrak before it leaves the area along the Norwegian coast. a minor part of the water which enters north of Scotland passes to the south along the British coast and fur-

SURFACE CURRENTS IN THE NORTH SEA AND SKAGERRAK

SALINITY OF SURFACE WATER PSU
- >35
- 34–35
- 33–34
- 32–33
- 31–32
- <31

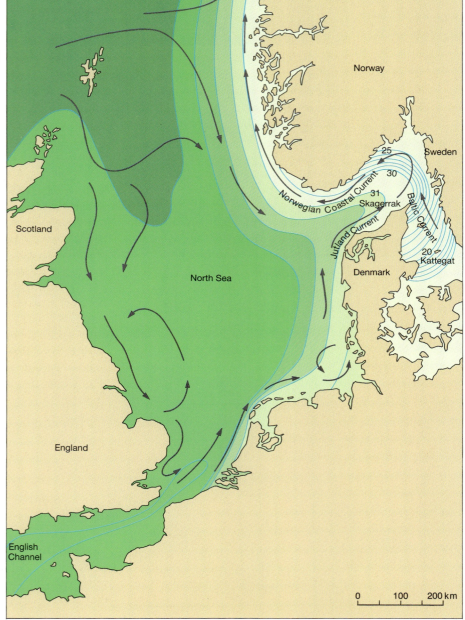

The Skagerrak's coastal regions and the Kattegat form the border between the oceanic North Sea and the strongly freshwater-influenced Baltic Sea. (G20)

TIDE IN THE ATLANTIC OCEAN

Amplitude at spring flood
- >6 m
- 4–6 m
- 2–4 m
- <2 m
- ○ Amphidromic point
- ⌇12 Co-tidal line

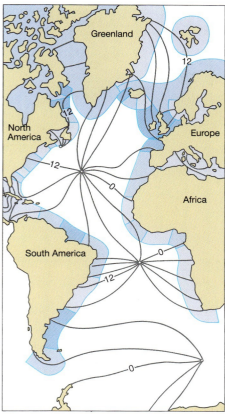

Highwater occurs simultaneously along a cotidal line. The figures show the time for highwater in "moon-hours" (ca. 62 minutes) after the moon has passed the 0° meridian (Greenwich). At amphidromic points the water level remains uninfluenced by the tide. The amplitude of the tide increases from these points towards the continents.(G21)

TIDE IN THE NORTH SEA

Amplitude at spring flood
- >6 m
- 4–6 m
- 2–4 m
- <2 m
- ○ Amphidromic point
- ⌇12 Co-tidal line

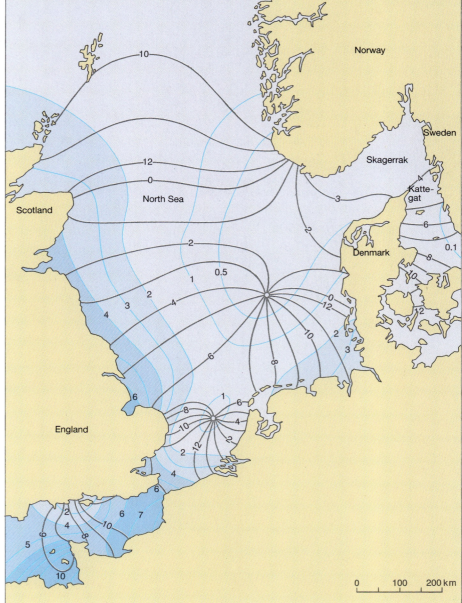

Tidal waves enter from the Atlantic through the English Channel and north of the British Isles. The cotidal lines show how the waves move anticlockwise around the North Sea.(G22)

ther to the east along the Dutch and German coasts, and then to the north towards the Skagerrak in the Jutland Current. Water from rivers and from the English Channel is added to the main flow on its way around the North Sea. In the Skagerrak, it joins up with water from the Kattegat and from rivers, mainly in Norway. The surface water in the Kattegat consists of one-third fresh water coming from the Baltic, and two-thirds water which earlier flowed into the Kattegat from the Skagerrak and became mixed with the Baltic water flowing out. The salinity distribution on the surface reveals how water with low salinity, follows the coasts around the Skagerrak.

Variations in Water Level

The water level at the coasts varies from hour to hour, from day to day, and throughout the year. These variations depend on the sun and the moon (tide-water), on wind and on air pressure, and are associated with the fact that the water surface slopes over an area of sea. Low water at one place corresponds to high water at another. The tidal forces cause slopes in such a way that tidal waves attain heights of one metre or more at the edges of the oceans. In the neighbouring Baltic, the same slope gives rise to a hardly noticeable tide of a few centimetres. The force of the wind against the water surface causes slopes which become greater the shallower the area of the sea. In shallow areas, such as the Baltic, the wind has a dominating influence on the water level. a strong wind which blows from the southwest gives an increase of one metre or more in water level at the head of the Gulf of Finland. The small annual variations observed in the water level of the Baltic, a decrease of a few decimetres in the spring and a small increase during the autumn, largely depend on differences in air pressure between the Baltic and the North Atlantic.

Water level variations generated in an area spread like long waves to other areas: The tides in the North Sea depend on tidal waves which come in from the Atlantic, mainly to the north of the British Isles. The waves increase in height in the shallow area of the sea. Owing to the effect of the Coriolis force on the tidal currents, the largest effect will be on the right-hand side, i.e., along the British North Sea coast, where the range may exceed 5 m in some places. The tidal wave migrates around the North Sea whereupon it is reduced as a result of friction against the bottom. At the opening to the Skagerrak the height of the wave has decreased to a few decimetres. The highest tidal range in the world, maximally more than 20 m, is found in Fundy Bay on the landward side of Nova Scotia on the Canadian Atlantic coast.

In addition to the tide, the wind causes major variations in water level in the North Sea. These variations are then transmitted to the Swedish west coast and cause the large variations in water level which sometimes occur in that area.

Elements in the Water – Mixing and Dispersion

For many years the sea was regarded as being infinitely large. As a result, we used the sea as a gigantic garbage dump in the belief that all waste would disperse and disappear in its depths. The effect this has had and is having on the sea and on the animal life it contains can only be guessed today. One example of how we have succeeded in influencing the living conditions in the sea is the decimation of the seal population in the Baltic, which is considered to depend on the seals being poisoned by PCB (a substance introduced by man).

Different substances have very different dispersal patterns. The spreading and distribution of different elements in the sea depends on factors which cause or restrict mixing in the water, as well as the physical and chemical properties of the element concerned.

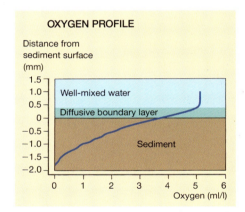

Molecular diffusion. Oxygen diffuses into the sediment and is consumed by degradation of organic matter.

TRACE ELEMENTS

Water is an extremely effective solvent, and it is believed that all elements in some chemical form are water-soluble to some extent. Consequently, sea-water probably contains all elements even though the solubility of a certain element is so low that we have so far been unable, using modern analytical methods, to detect all of them (62 of the 89 naturally occurring elements have been found in sea-water). Only 14 of the elements occur at concentration levels higher than 1 mg/l. Other elements, with concentrations below 1 mg/l, are called trace elements. This definition also includes the nutrients, i.e., phosphorus, nitrogen and silica compounds. Certain other trace elements in sea-water are just as important as the nutrients for marine life, e.g., certain metals, whereas others such as radioactive elements, may imply a threat.

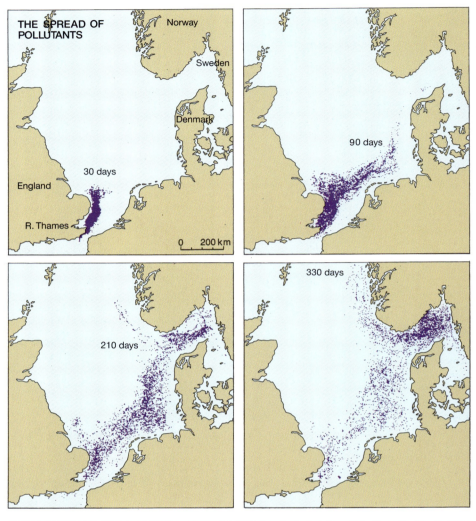

Turbulent diffusion. The example shows a simulated discharge of passive particles from the mouth of the river Thames and its dispersal throughout the North Sea.

Water can be transported for long distances without its characteristics being changed. Transport of the Baltic water (light brown) up through the Kattegat. (7 May 1990, NOAA-AVHARR)

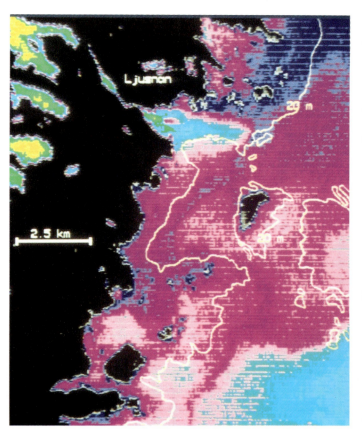

Satellite pictures can be used to study how supplies from rivers disperse into the sea. The photo shows the dispersion area of the river Ljusnan, 2–3 km off the coast in Ljusnefjärden. (11 October 1988, LANDSAT)

At work in a clean-air laboratory. SWEDARP 88/89, a Swedish expedition to the Antarctic.

Water sampling from R/V Argos, operated by the Swedish National Board of Fisheries.

What is Mixing?

Physically, mixing may be described as a transfer of properties (e.g., heat, concentrations of different elements). An example is when the contents of a coffee cup are stirred in order to mix the coffee with milk or sugar. The cause of the mixing in this case is the movement of the spoon, which leads to the formation of eddies, turbulence, which transfers the properties of the coffee to the milk and sugar, and vice versa. Almost all mixing in the sea takes place by the formation of eddies, turbulent diffusion. The efficiency of the mixing, e.g., the magnitude of the dispersal, will be extremely dependent on the size of the eddies. The smallest eddies, which may be 1 mm or smaller, are isotropic, the same size in all directions. The largest eddies are so large (10–100 km) that they are influenced by the stratification, topography and sometimes also by the Earth's rotation. They are also frequently strongly compressed vertically and are then called anisotropic. On very small scales, such as in the border zone between water and air, or between water and sediment, molecular diffusion may be of major importance since it controls the flow of different substances straight through these layers. Molecular diffusion is much slower than turbulent diffusion. In the example with coffee and sugar, it would require weeks for the molecular diffusion alone completely to mix the sugar in the cup.

CAUSES AND LIMITATIONS OF MIXING

There are many factors which cause mixing in the sea. Wind and waves are perhaps the most obvious factors but also currents which vary in strength in time and space cause mixing. Under certain conditions, a coastal current, for example, may become instable and large anisotropic eddies are formed which carry properties/elements out into the open water. Instead, when the current is stable, mixing is reduced and water in the current can retain its identity over long distances. Such conditions sometimes occur on the Swedish west coast where periodically the brackish Baltic water penetrates as far as up to northern Bohuslän.

Cooling/heating may also cause mixing; particularly in marine areas which have a temperature stratifica-

ORGANIC ELEMENTS

Sea-water contains large amounts of naturally produced organic substances which the living organisms need for growth, and which are excreted from them as degradation products. Among these we can find amino acids, carbohydrates, fats and also residues of decomposed plants – generally called humic substances. The amount of dissolved organic material in seawater is many times greater than that bound in living animals and plants.

In addition to natural organic substances, mankind has added a large number of industrial products, of which some have a damaging effect on the environment. Polychlorinated biphenyls (PCB), dioxins and certain oil components (PAH) are examples of toxic organic substances which have been discharged into the sea as a result of industrial activity. There is still lack of knowledge on these substances, and today, for example, we can only identify a low percentage of all chlorinated substances in industrial emissions.

Other substances, e.g., chloroform (formed by chlorination of water in sewage and power plants or when bleaching paper pulp), perchloroethylene (earlier used as a solvent and cleaning fluid) and chlorofluorocarbons (used as spray gases, in refrigerators and as cleaning fluid in the electronics industry), can be found in relatively high concentrations in seawater but we do not know whether they have a toxic influence on life in the sea.

These products of industrialism have the property that they are more or less volatile, which implies that as soon as they get in contact with air they evaporate to the atmosphere and are carried long distances by the wind. Since the atmosphere, when considered in a global perspective, is mixed much faster than the sea, this has the effect that, for example, PCBs, have been found in animals caught in the Antarctic despite most of the sources being located in the northern hemisphere. In the atmosphere, as a result of solar radiation and certain radicals, there is a slow degradation of the various organic substances. In the sea-water, on the other hand, they are comparatively stable and persistent, i.e., resistant to different chemical and biological degradation processes, and may remain for long periods before they are degraded to less hazardous substances. They are also fat-soluble, which means that they accumulate in the fatty tissues of living organisms in the sea such as seals and birds.

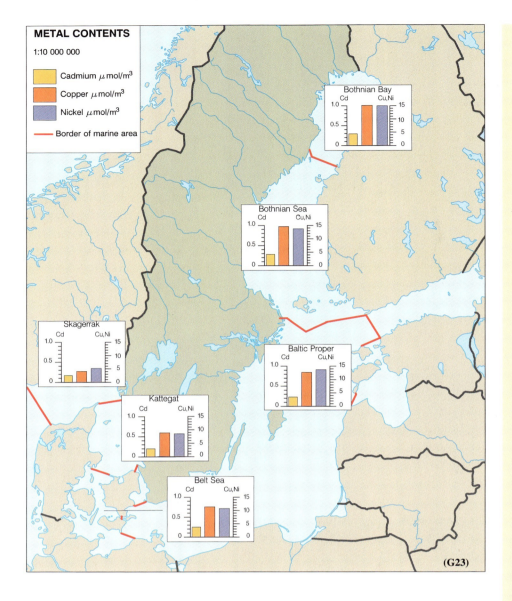

METALS

Some of the most important trace elements are metals, e.g., iron, copper, lead and magnesium which, as chemical compounds, mainly occur bound to particles or dissolved in sea-water. They are discharged into the sea with runoff from land and by precipitation. Metal concentrations in sea-water vary widely from area to area, partly depending on the different amounts of trace elements in the fresh water entering the sea. The concentrations may also be dependent on the oxygen content of the water. In the Baltic, for example, with a large supply of fresh water and varying oxygen conditions, the metal concentrations may diverge considerably from those considered normal in the open sea, and the variations may also be large within the Baltic itself. Trace metals also flow into the sea in emissions from industrial and other human activities. Emission of exhaust gases from traffic and the use of studded tyres lead to large amounts of heavy metals entering the sea, particularly during the snowmelt season in the early spring.

Some trace metals are of major biological importance, e.g., iron is the central atom in haemoglobin, magnesium in chlorophyll, and cobalt in the growth hormone B12. Although these elements are essential to life, they may be toxic in high concentrations.

Many more or less insoluble metals accumulate in different organisms and, consequently, their concentrations are often magnified in the upper levels of the nutrient chain. In organisms at the top of the chain, e.g., seals in the Baltic, the metal concentrations may be 1,000–10,000 times higher than in the ambient water. Since trace metals frequently occur bound to particles, there are also considerable amounts accumulated in the sediment.

Knowledge of the occurrence of trace metals in sea-water and their biological importance is still very deficient. Only recently has it been possible to make reliable measurements of metals to any greater extent in waters around Swedish coasts.

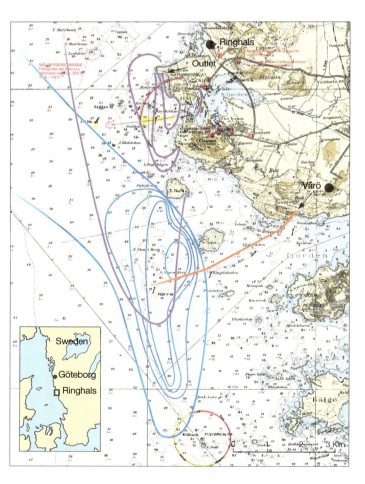

tion. In the Bothnian Bay, for example, vertical mixing takes place each spring and autumn when the water is heated (during the spring) or cooled (during the autumn) and the temperature approaches the temperature for maximum density. Surface water will then become heavier and sink. If the water mass is salinity stratified, this will be prevented, as in the case of, for example, the Kattegat where the exchange across the halocline is strongly restricted.

Stratification may restrict vertical turbulent exchange, i.e., the vertical size of the eddies. In stratified water, an emission outflow onto the surface may, under calm conditions, have a considerable horizontal extent. Topography and varying driving forces

The depth at which discharges occur may be decisive for their dispersal. Discharges from the Ringhals nuclear power station (violet) are made into the surface water, whereas those from Värö bruk (orange) are through an outlet tube below the halocline.

may also cause very large eddies, so large that their movement is influenced by the Earth's rotation. In these eddies, the mixture may be effective but they may also be enclosed and transport large masses of water over large distances, since they can persist for several weeks. The inflows to the deeps of the Baltic appear to take place in this way.

DIFFERENT ELEMENTS ARE MIXED TO DIFFERENT EXTENTS

A conservative element such as salt, i.e., an element which is dissolved in water but in other respects is not influenced chemically or biologically, is mixed passively with water and is influenced only by dilution. A different situation is found with elements which interact with their surroundings, and the mixing processes and ways of dispersal for these elements are much more complicated.

In connection with the Chernobyl accident in 1986, the coast of Norrland received a massive dose of cesium−137 (Cs137). Most of this was deposited on land where it rapidly became bound to particles of different size, carrier particles, and was taken up by plants and animals. Subsequently, the continued dispersal of the substance was largely governed by the dynamics of the larger particles. If the particle-bound material enters a lake or a bay of the sea it will mainly be agglomerated into larger particles and will rapidly sink to the bottom. Consequently, most of the fallout Cs137 can be found on the sea floors in coastal areas and only low concentrations can be measured in sea-water. Owing to the variation in the size of the carrier particles, fine particles will remain in suspension longer than coarse particles. This will imply that small particles can be transported over long distances whereas larger particles remain in the neighbourhood of the source.

Substances lighter than water, such as oil, will naturally disperse differently to material which sinks to the sediment. As regards substances which remain on the surface, the dispersal is dominated by the prevailing wind and the surface current.

WHEN IS THE MIXING MOST EFFECTIVE?

The efficiency of the dispersal of a discharged substance depends on the turbulent length scale, i.e., the size of the eddies, and the relationship between the turbulent length scale and the size of the discharge. If we look at a "cloud" of a discharged substance, we find that the cloud is mixed to an extremely small extent if it is enclosed in an eddy which is considerably larger than the size of the discharge. Instead, the cloud will passively follow the large-scale current. Since the turbulent length scale is small in comparison with the size of the cloud, almost all mixing takes place within the cloud without really achieving any dispersal. Only gradients in the actual discharge are levelled-out. The most effective dispersal takes place when the size of the eddies coincides with those of the cloud. An empirical relationship has been found between the size of a discharge and its dispersal. The larger the "cloud" the faster the mixing, since larger eddies will then have an influence.

RADIOACTIVE ELEMENTS

Sea-water has a certain background radioactivity, mainly originating from the radioactive potassium isotope ^{40}K. Some of the naturally occurring radioactive elements have been present since the Earth was created, others originate from the atmosphere where they are formed more or less continuously (by cosmic, energy-rich radiation). Since 1945, as a result of nuclear bomb detonations, the nuclear power industry, etc., the radioactive radiation of sea-water has increased to a measurable degree. Earlier, radioactive substances were emitted into the Baltic as a result of discharges and dumping without any real control. Today, this has largely ceased and the control imposed by different international and national agencies is rigorous.

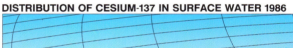
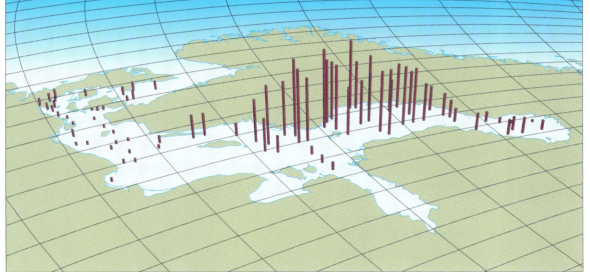

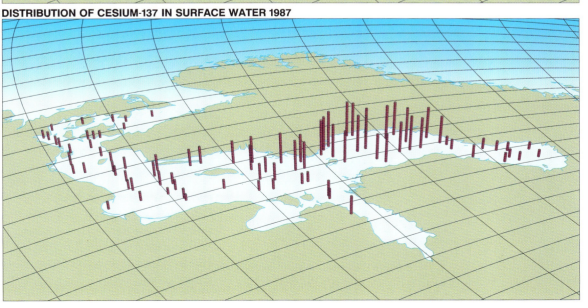

Distribution of Cs137 (20 Bq/m³ per scale unit) in the surface water. For many years after the Chernobyl accident, the concentration of Cs137 will remain high, mainly in the Bothnian Sea, as a result of leaching from land.(G24, G25)

Water in the West and Water in the East

Water Turnover, Fresh Water Supply and Salinity

The large-scale circulation in the seas around Sweden may be said to depend on three factors: Bottom conditions, water exchange with the North Sea, and the supply of fresh water. Schematically, the circulation may be said to consist of a surface current with brackish water which flows out from the Baltic through the Kattegat to the Skagerrak and the North Sea, and a current in deeper layers with more saline water which goes in the opposite direction. The boundary layer between the brackish and the salt water, the halocline, may be more or less sharp. The fresh water largely comes from rivers in the drainage area of the Baltic, an area which is several times larger than the sea itself. Direct precipitation on the sea surface is almost entirely balanced during the year by evaporation. The entire Baltic is influenced by this inflow of fresh water. The surface water in the Baltic Proper, and the water in the Bothnian Sea and Bothnian Bay, consists of more than four-fifths of fresh water and only one-fifth of saline Skagerrak water.

The inflow of fresh water to the Baltic may be described as the engine which drives the large-scale circulation. The inflow generally causes a higher water level in the Baltic than in the Kattegat and Skagerrak. This difference in water level forces the brackish surface water out of the Baltic. On its way towards the Skagerrak the brackish water becomes increasingly saline since the surface water becomes mixed with the underlying water. In order to replace the water entrained into the surface current, an undercurrent of more saline water is formed which runs through the Kattegat and further down into the deeps of the Baltic.

The salinities in the Skagerrak are almost as high as in the North Sea, and Baltic water can be traced only in a thin surface layer along the coasts of the eastern and northern Skagerrak. The Kattegat is considerably more influenced by the out-flow from the Baltic. the Sound, the Belt Sea and Kattegat together make up the shallow sill area which restricts the exchange of water between the Baltic and the Skagerrak. It is here that most of the mixing takes place between Baltic water and the saltier water from the Skagerrak. Usually, there is a halocline that covers the entire Kattegat. The lowest salinities are found in the surface farthest to the south, and the highest in the deep water farthest to the north. By entrainment of water from below, the salinity in the surface increases towards the north. During the winter, when wind and cooling interact, the mixing may be extremely effective.

Having passed the shallow sills, the saltier, heavier Kattegat water spreads out into the Baltic close to the bottom along a number of channels. At the same time, it becomes mixed with surrounding brackish water and increases in volume. Mainly as a result of mixing by the wind, the surface layer has an almost homogeneous salinity. A halocline separates this water from the more continuously stratified deep water. The halocline limits the turbulence and decreases the vertical mixing. Finally, the inflowing water is stratified at the level which corresponds to the density obtained by the bottom current as a result of it becoming mixed with the surrounding water. Normally, stratification takes place around the halocline but sometimes, when meteorological factors are favourable, large inflows of salt Kattegat water take place which are able to displace the water masses present in the deepest basins. Between these events, salinity slowly decreases in deeper areas as a result of vertical mixing; these periods are usually called stagnation periods and have a strong influence on the oxygen and nutrient situation in the Baltic.

The deep water of the Bothnian Sea originates from the surface water of the Baltic Proper. Strong mixing and large supplies of fresh water ensure that the salinity stratification of the Bothnian Sea is weak. In the Both-

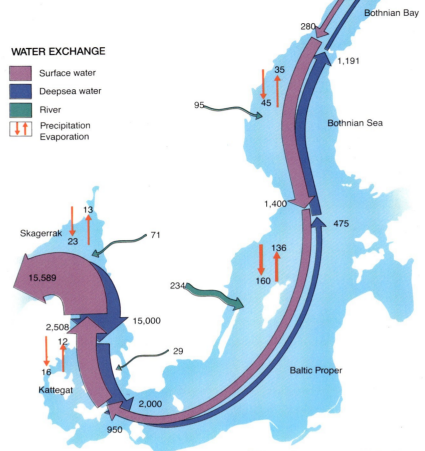

Water exchange expressed in km³/year. (G26)

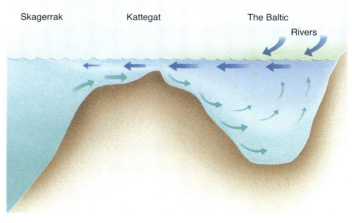

LARGE-SCALE WATER CIRCULATION

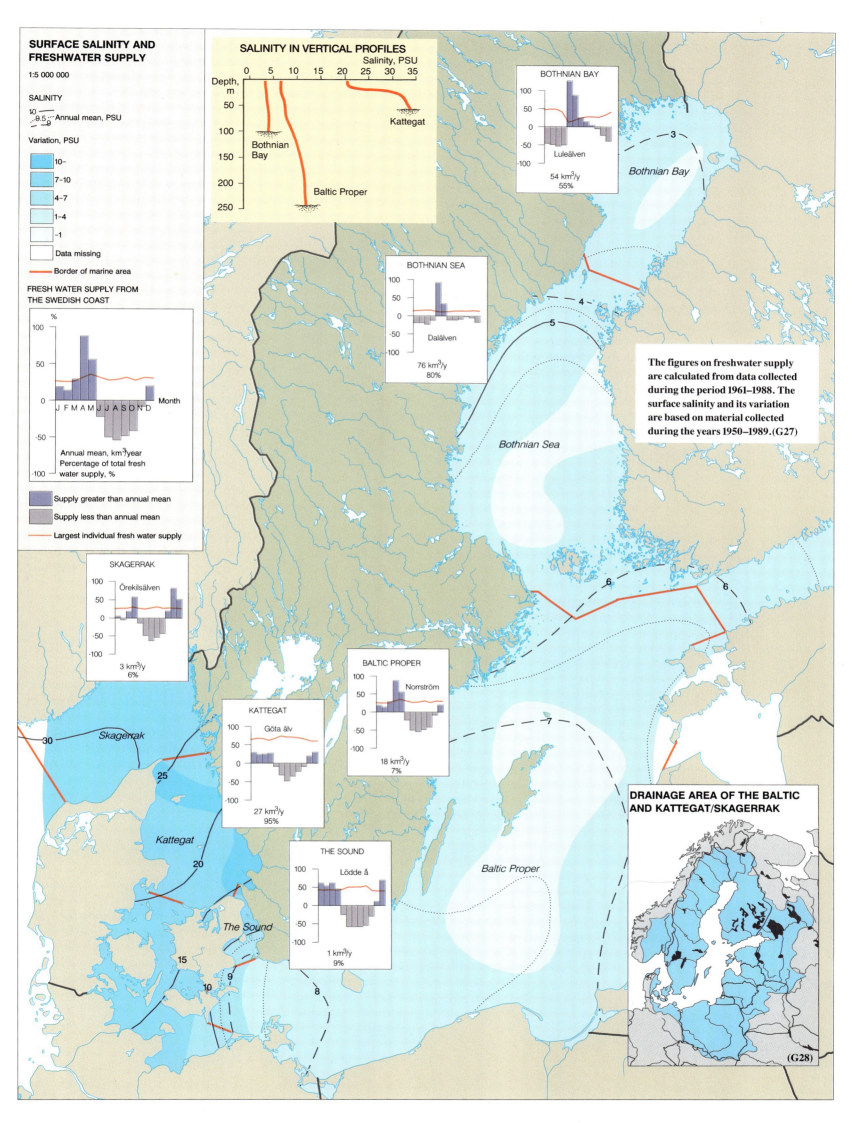

nian Bay the salinity has decreased down to only a few units (PSU).

The rate of turnover of water in the various basins can be estimated by calculating the flows between different sub-basins on the basis of volume and salt balance. The water of the Bothnian Bay has a turnover time of about 4–5 years whereas 25–35 years is required for turnover of the entire water mass in the Baltic. Thus, pollutants entering the Baltic will remain in the system for a long time. In the Kattegat and Skagerrak, turnover times amount to some months, whereas water in the narrow Danish sounds is replaced within a matter of weeks or sometimes within a few days.

Temperature and Ice

During the spring, the sea is heated by solar radiation and by air temperature being higher than water temperature. When the autumn arrives, the surface of the sea is, on the other hand, warmer than the air and thus the water instead becomes cooler. The cooling continues during the winter and ice is formed, which protects the water from further cooling. As a result of the large variations in weather and wind throughout the year, there are changes in the temperatures of the surface water around the Swedish coast from winter temperatures close to zero up to summer temperatures in excess of +20°.

When the surface water is heated, a thermocline is formed. The warmer water is slightly lighter than the deeper water and may thus insulate the deeper layers from the surface. During the spring, the thermocline is close to the surface but its position becomes deeper during the summer as a result of mixing. In the southern Baltic it usually reaches down to a depth of 30 m. Mixing of the surface water largely depends on the wind. Severe autumn storms may mix the surface water all the way down to the halocline. The thermocline is disintegrated and is not reestablished until the following spring when solar radiation again heats up the surface water.

As a result of this vertical mixing, colder water from deeper layers is mixed with the warmer surface water. Thus, in a year with strong vertical temperature stratification, a storm may rapidly alter the temperature of the surface water. Vertical mixing may also take place since the surface water will become heavier as a result of heating or cooling. This may occur during the early spring when the water is warmed up to the temperature for density maximum, or during the autumn when the water is cooled to the same temperature. The mixing caused by thermal effects, however, only occurs within layers where the salinity is constant, e.g., down to the halocline. This is because the salinity stratification effectively reduces the mixing.

Another example of how the temperature of surface water can rapidly change is when situations occur with upwelling. The surface water is then blown away from the coast and replaced with colder deep water. This phenomenon is common along the Swedish east coast when the wind is blowing from the south and southwest and may cause temperature changes of 4–5°C in only a few hours.

During the winter, the surface water is cooled so much that ice may form. However, the extent of the ice varies widely from year to year depending on mild or cold weather. The first ice starts to appear in the innermost bays of the Bothnian Bay during mid-November. During a normal winter, the entire Bothnian Sea, Sea of Åland, Gulf of Finland and the northernmost part of the Baltic are also covered by ice. Similarly, there is ice-cover in the archipelagos of the Baltic, the Kalmarsund and Bohus archi-

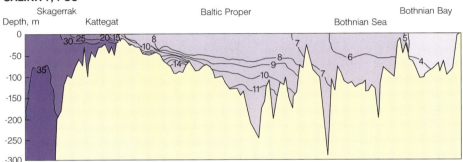

Distribution of salinity along a section from the Skagerrak to the Bothnian Bay, 1988.

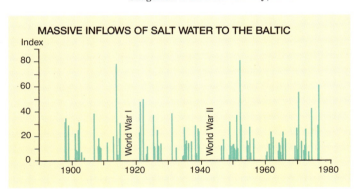

The size of the inflows is characterized by an index (Q) based on duration and mean salinity. High values represent large inflows.

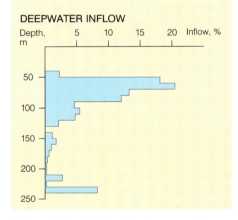

Estimated distribution of the deep water flow from the Kattegat into the Baltic Sea. Only occasionally are the inflows of such an extent that the deepest parts of the Baltic are affected.

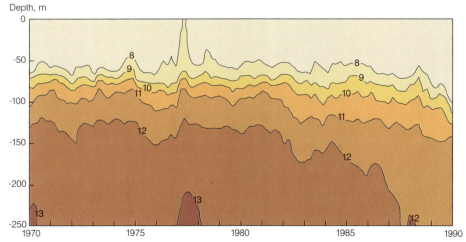

Salinity concentrations in the Central Baltic Sea have decreased in deep water since 1976.

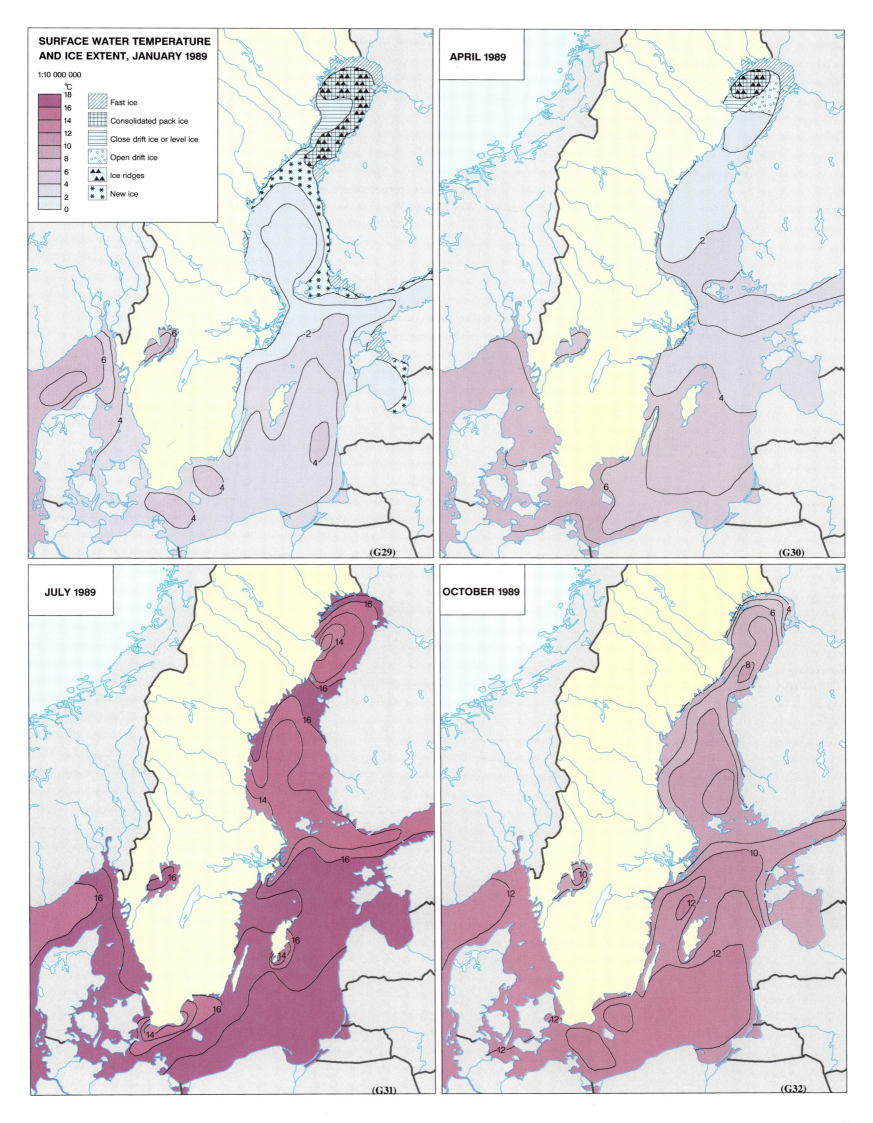

Trawlers in a frozen sea, January 1987. The Sound freezes only when winter conditions are severe.

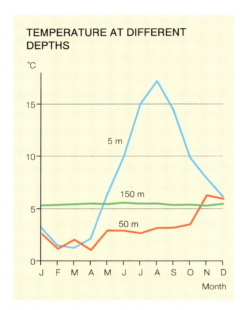

TEMPERATURE AT DIFFERENT DEPTHS

Temperature variations in central parts of the Baltic Sea. Monthly mean values for 1980–1990.

Temperature pattern from the Quark and parts of the Bothnian Bay as measured by satellite. Southwesterly winds have caused upwelling along the coast. The temperatures vary between 7 and 10°C, the coldest water being along the coast. (6 October 1981, NOAA–7)

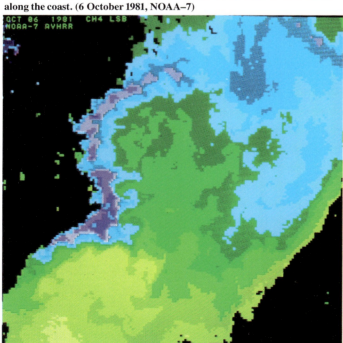

pelagos. During severe ice-winters, also the Kattegat/Skagerrak, the Belt Sea, the Sound and large parts of the Baltic are covered by ice. However, it is very unusual for the entire Baltic to have ice-cover at the same time. The maximum ice-cover in 1987 as interpreted from satellite pictures showed, for example, that about 90% of the Baltic was covered by ice. A corresponding ice situation might have occurred during the severe winter of 1942.

Several different types of ice are found at sea. As the sea is never absolutely calm, the ice moves with the currents and the winds and becomes compacted mainly along the coasts. That part of the ridge above the water surface is usually called the sail whereas the part in the water is called the keel. The keel may be several metres deep and up to 28 m deep ice ridges have been measured in the Bothnian Bay. Waves may cause slush and shuga to form pancake ice. The rubbing of the ice floes against each other gives them an upright edge of shuga and a rounded shape. Old ice, which is melting, is usually called rotten ice. In general, this is dark and porous and the ice crystals easily break when touched.

SUMMER TEMPERATURE

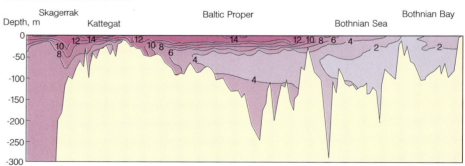

WINTER TEMPERATURE

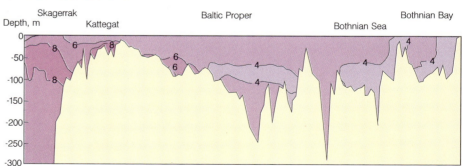

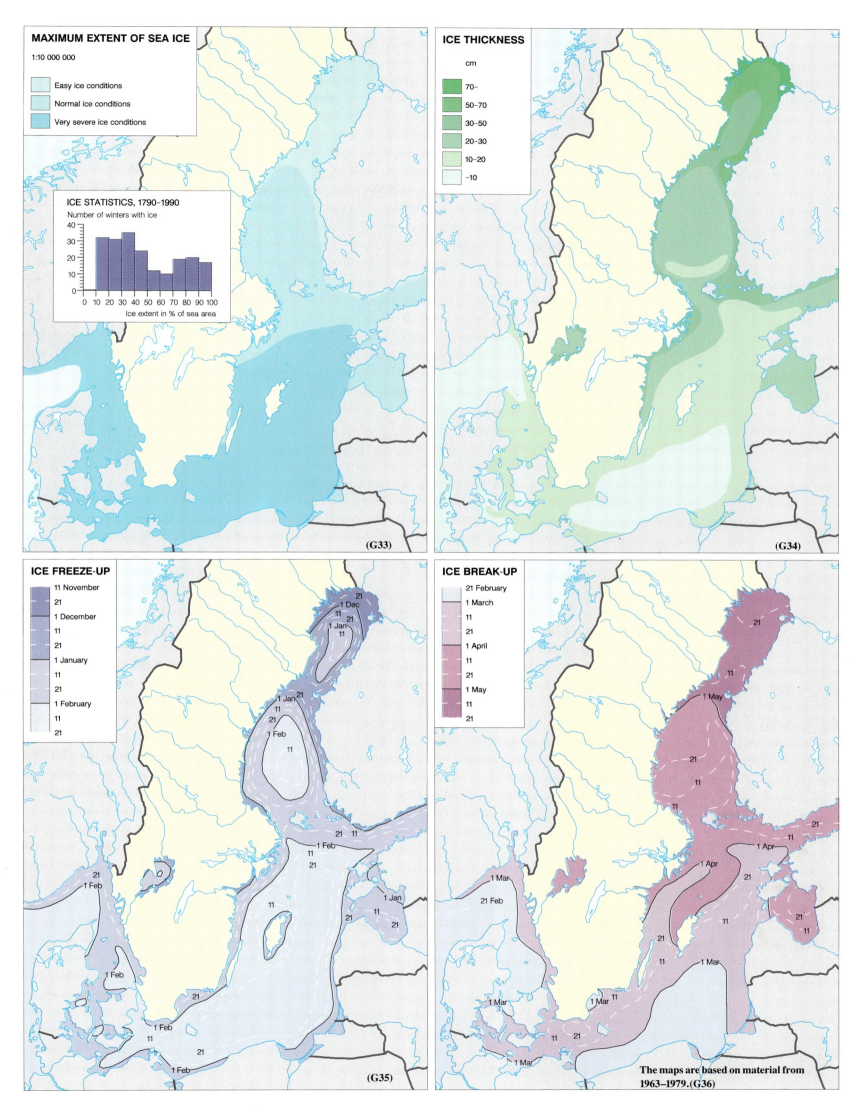

The diagram is compiled from observations of wave climate at Ölands södra udde. The significant wave height is calculated from 10-minute observations and comprises the mean of the highest one-third of the waves during the period. Glommen lighthouse to the north of Falkenberg.

Wind Waves

Waves are formed when the wind blows over a surface of water. At wind velocities of about 0.5 m/s the water surface becomes covered by small ripples which cause the sea to change colour, from shiny grey to azur blue. As the wind increases, the waves become higher and longer. Energy is transferred from the wind to the movement of the wave. The waves transport this energy further but they do not transport any water.

If the wind suddenly increases to 10 m/s it will require 6–8 hours before the waves field has become fully developed (saturated). A wind of hurricane strength, 32 m/s, must blow for more than 24 hours in order to generate a fully developed sea.

The wind must also blow over a certain minimum distance (the fetch) in order to generate a saturated wave field. Wind velocities of 10 m/s require a distance to the lee coast of about 150 km and at velocities of 32 m/s an open stretch of 1,000 km is required.

Thus, the Kattegat is just sufficiently large for a wind force of 10 m/s to give a saturated sea. Hurricane-strength winds occur extremely rarely for such a long period as an entire day, but if this were to occur in the Baltic the blowing distance is still insufficient to cause fully developed hurricane waves.

During the initial stage, or in the area of initiation, the waves are short and steep and commonly break, i.e., a choppy sea. The breaking waves transfer energy to waves with longer wave lengths and the wave field gradually becomes dominated by increasingly longer waves. Energy is also transferred by breaking waves to shorter wave lengths which are damped through friction, thereby causing the kinetic energy gradually to be transferred into heat. When equilibrium is reached in the flows of energy so that all energy supplied by the wind is transferred to short waves and thermal energy, the wave field is no longer modified. The long waves which have developed are in equilibrium with the prevailing wind velocity. When this occurs, waves in the open sea do not break as frequently as during the initial stage. The reason for breaking waves during a period when the wave field and the wind are in equilibrium is interference – interaction between different wave groups with slightly different wave lengths or direction.

When the wind decreases the equilibrium is altered and energy leaves the wave field, mainly by the shorter waves being reduced. The long uniform waves, swell, will remain. The swell often persists for some days after the wind has abated. If the wind turns to another direction and creates

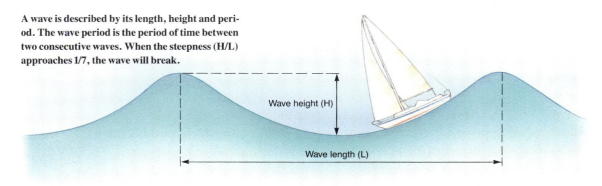

A wave is described by its length, height and period. The wave period is the period of time between two consecutive waves. When the steepness (H/L) approaches 1/7, the wave will break.

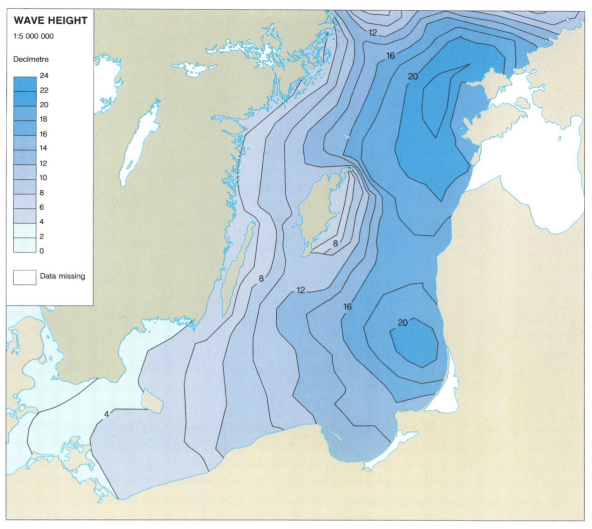

The forecast map for midnight between 23–24 August 1990. The map shows wave heights after 24 hrs, with W-NW winds 7–10 m/s, in the northern and central Baltic Sea. The highest wave heights are generated in places with the largest fetch from the NW, e.g., off Estonia and Latvia. To the east of Gotland the wave height is small because of the lee effect arising during off-shore winds.(G37)

Internal waves can sometimes be seen as streaks on the surface. Hårsfjärden in the Stockholm archipelago.

SHORT AND LONG WAVES

In deep water, short waves move at a phase velocity of $c=(gL/2\pi)^{1/2}$ or $c=gT/2\pi$ ($\pi=3.14$ and $g=9.81$). The water particles move in circular orbits with diameter H with velocity of $\pi H/T$. If T=5 sec and H=0.5 m then the wave velocity will be 7.8 m/s and the particle velocity will be 0.3 m/s.

When the movement of the wave is disturbed by the seafloor, which occurs if the depth is less than half of the wave length, the particle movements become increasingly oval. In shallow water, particle movement is almost entirely forwards and backwards (swell). The phase velocity then no longer follows the formula above, instead the velocity is $c=(gd)^{1/2}$, where d=water depth and the waves are said to be long.

a new wave field which interacts with the old swell, then difficult sea conditions may occur.

Waves may sometimes be reflected against a steep rocky coast where the depth is so great that the wave can roll in without losing its energy through friction against the bottom. Storm waves from the southwest which roll in towards Islandsberg north of Gullholmen or Tjurpannan at Havstenssund in Bohuslän are reflected out again towards the southwest with almost unaltered wave height, where they meet the incoming waves. This creates a steep-crested wave field, dangerous to navigate.

Waves in interaction with currents, and particularly tidal currents, are famous for their turbulence in, for example the English Channel, at the Orkney Islands or at Lofoten.

WAVE HEIGHT

A wave field consists of numerous waves of different length and height. Thus it is difficult to talk about a general wave height. In order to estimate wave height from, e.g., a lightship, it has been agreed to state the mean height of the highest one-third of the waves ⅓ measured from crest to trough.

Extreme wave heights for Swedish conditions have been measured during the winter months when the wave height $H_{1/3}$ has occasionally been about 8 m at Öland's Södra Grund and at Almagrundet (January). At Trubaduren, off Göteborg, a wave height of almost 6 m (October) has been measured. The highest individual waves on these occasions were higher than 10 m.

In comparison, it may be mentioned that during a windy winter (1968) in the North Sea the wave height was more than 3 m for almost 50% of the time. The highest measured wave during this winter was 18.6 m.

CALCULATIONS OF WAVES

The large computers in which the meteorological forecasts are prepared can also be used to calculate wave height in, e.g., the Baltic. The wind forecast is entered into the calculation system together with data on water depth and coastlines. The computer calculates the energy transfer from wind to water surface, the transport of wave energy (swell), the energy loss when the waves enter shallow water, and subsequently wave height and direction in the other parts of the sea.

OTHER TYPES OF WAVES

The discussion above has dealt only with types of waves which are generated by wind, but there are also other types of waves in the sea. Two examples are tidal waves caused by the force of gravity in the solar system, and tsunamis which are caused by earth tremors, mainly below the water surface. All these types of waves have the common factor that they have very long wave length, thousands of kilometres, and they are experienced more as periodical changes in the water level rather than as waves.

Not only can the sea surface move up and down, but waves also occur deep in the sea. If the water is stratified with, e.g., one layer of fresher water above more saline water, then wave movements may occur in the border zone between these layers, so-called internal waves. When these waves break, they may be of great importance for the mixing in the sea.

Water Level

Interest in water level (i.e., the position of the water surface) and the variations in the sea's water level can be traced back many years. The practical importance of water level for shipping and the building of harbours was recognized early. In Sweden, the first regular measurements of water level started in the late 18th century. In 1774 the Government of Gustav III issued directives dealing with the measurements of water level in Stockholm, where both the water level in Lake Mälaren and in the Baltic were to be registered. This directive was the result of complaints from representatives of shipping interests with regard to difficulties in passing vessels through the locks linking Lake Mälaren and the Baltic owing to the variable differences in water level between these areas. The water level is still being measured in Stockholm and today this series of measurements is the second longest in the world, the measurements of water level in Amsterdam being started earlier.

In general, the water level is given in relation to the mean water level. This is not fixed but changes, mainly on account of land uplift (isostatic changes) which is largest in the north. In the far south of Sweden, land uplift has been replaced by subsidence. As a result of the melting of the glaciers, the mean water level of the global ocean is increasing (eustatic changes) but this effect is considerably less than the effect of land uplift. The annual mean water level is a term used in Swedish sea reports and is a calculated mean water level from which the isostatic and eustatic changes have been subtracted.

In Swedish waters it is mainly the varying wind conditions that regulate the water level around the coasts. On the west coast, for example, the water level may frequently increase 1 m or more during periods with strong westerly winds. Air pressure also influences the water level; a high pressure over the Baltic will force water out through the Danish sounds and the level in the Baltic will decrease. The water level, mainly in the Baltic, is also influenced to some degree by fluctuations in the fresh water supply. The effect of tidal water on Swedish seas is small in comparison with other effects.

Water level varies most during the autumn and winter when the winds are strongest. During spring and summer, when the weather is more stable, the variations are larger and are then determined mainly by air pressure.

The highest high water level, in relation to the mean water level, ever measured in Sweden is 181 cm measured at Kalix in the Bothnian Bay in 1984 and the lowest low water level, −144 cm, was measured at Ystad in 1902.

During strong W-SW winds the water level along the west coast rises, leading to flooding in exposed places. Grebbestad in Bohuslän.

PERIODICAL VARIATIONS IN WATER LEVEL

Normally the water level varies in accordance with variations in wind and air pressure. Particularly in the Baltic, variations in water level may occur with a period of about 27 hours, and are caused by a type of standing waves which occur when the wind subsides after a period of, e.g., southwesterly winds. Under such conditions, the northeastern Baltic and the Gulf of Finland will have an increased water level whereas in the southwestern parts the water level will be lower than normal. When the wind abates, the water attemps to level-out the differences. The water then flows to the south whereby the water level decreases in the north and increases in the south. The accelerations are so powerful that the situation will be completely reversed, i.e., high water level in the south and low in the north. When the highest water level has been reached in the south, the currents reverse and once again the water level will become high in the north and low in the south. In this way, the water level can alternate periodically in the north and south for about a week unless wind conditions change. These standing waves may cause serious flooding in the coastal towns of the Gulf of Finland. Similar waves are also found in lakes and are called seiches, a Swiss-French word which simply means variations in water level.

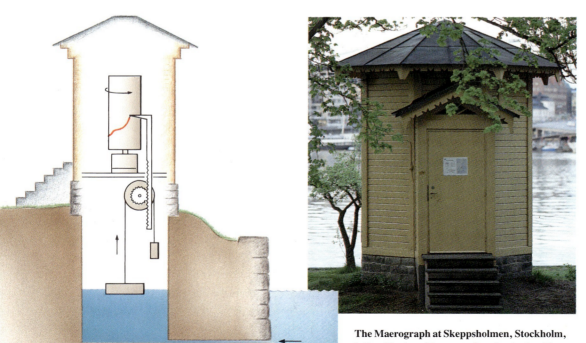

The Maerograph at Skeppsholmen, Stockholm, is the oldest working recorder of water level in Sweden. Similar instruments are located at about 20 places around the coast of Sweden.

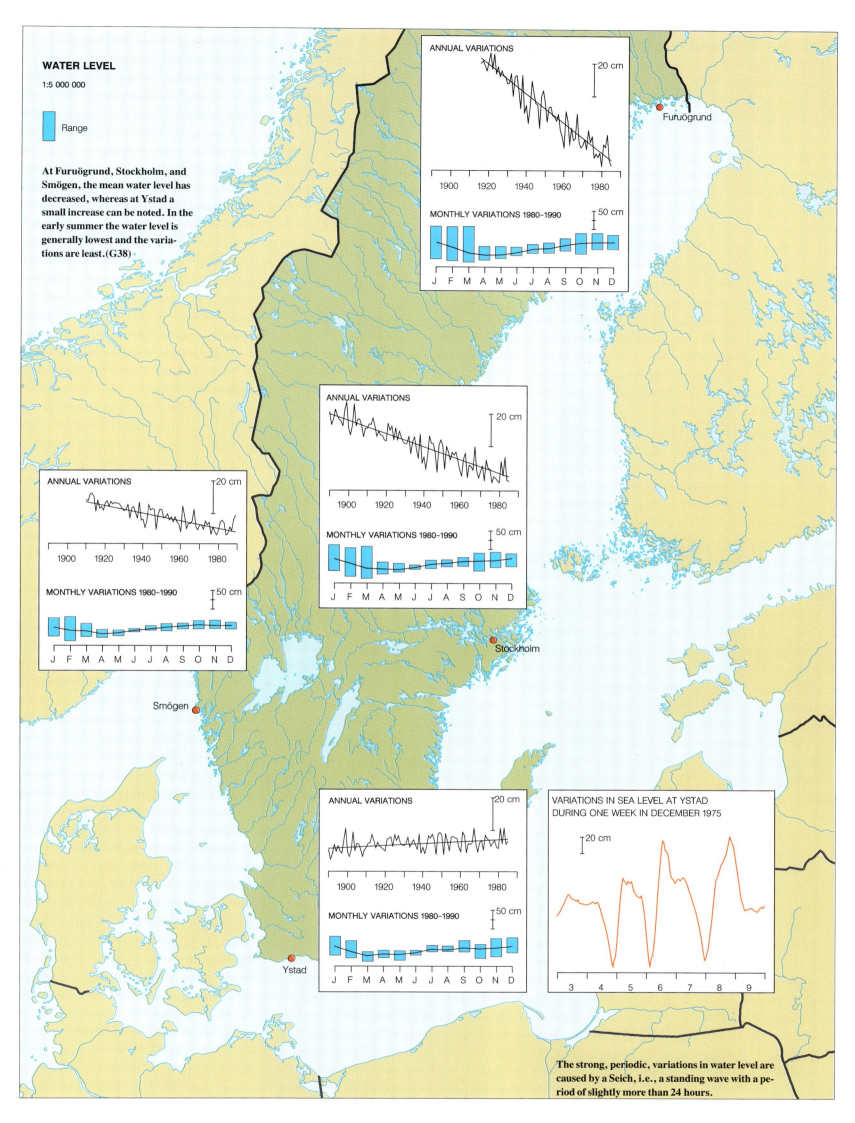

Currents

Knowledge of currents, their strength and direction, was first obtained during the era of the sailing-ships. The Jutland Current is said to be the first to be given its own name.

Currents in Swedish waters are mainly driven by wind and consequently vary strongly from one time to another and from place to place. The wind-driven currents reach speeds of 1–2% of the wind velocity and are directed to the right of the wind direction on account of the rotation of the Earth. Variations in water level caused by changes in air pressure and fresh water supply may also have an effect on currents.

Generally, there are several factors which cause a current, whereby different movements are superimposed on each other, giving the current an irregular and randomized appearance without any direct link to a mechanism involving driving forces. This has led to the properties of currents frequently being described in the form of mean values obtained from measurements made over longer periods. The character of the mean current depends on the length of the time period from which the mean value has been calculated.

On an annual basis, the mean current through the Sound is based on an outward transport of brackish surface water from the Baltic and an inward transport of more saline Kattegat water at deeper depths. Superimposed on this circulation is a current which is driven by the difference in water level between the Kattegat and the Baltic. The difference in water level varies depending on variations in the wind and air pressure. Despite the net transport through the Sound and the Belts being almost exclusively determined by the supply of fresh water to the Baltic, it is the currents driven by water level that dominate and which mainly govern the supply of oxygen-rich salt Kattegat water to the deeps of the Baltic.

The currents in the Skagerrak are the most durable in Swedish waters. The mean current direction is cyclonic, i.e., anti-clockwise, both at the surface and deeper. The surface circulation is driven by the supply of fresh water and the prevailing westerly winds, whereas the rotation of the Earth, the topography of the seabed and the coasts have a governing effect. The current at greater depths is more durable than the current at the surface and is controlled by the water exchange with the North Sea. It is assumed that the water entering the Skagerrak, particularly water from the southern North Sea, carries with it very large amounts of nutrients and environmental pollutants where they are partly deposited in the central parts.

The Kattegat and the Baltic lack a clear circulation pattern similar to that in the Skagerrak. The most important cause of currents in these two areas is the wind. Different wind directions cause completely different circulation patterns.

However, the supply of brackish water to the Kattegat from the Baltic means that the surface water in the Kattegat, on average, moves northwards. In situations with southwesterly winds, the brackish water forms a well-defined current which runs along the Swedish west coast to the north, the Baltic Current. In the Baltic, a weak cyclonic mean circulation can be distinguished with current velocities corresponding to about one-tenth (1–3 cm/s) of what is normally observed. The strength and direction of the mean current in the Baltic is determined by factors similar to those found in the Skagerrak but with a lower velocity and less durable.

ROTATING CURRENTS

Inertia currents often occur in the open sea. These are formed when the wind abates in strength or completely subsides. The efforts of the wind to move the water along its own direction are counteracted by the Coriolis force, an effect which works at right-angles to the current. When the wind ceases to blow there will be inertia in the water, expressed as an attempt to retain its movement. When the wind no longer has a controlling influence, the Coriolis force will cause the movements to diverge off towards the right; a form of pendulum movement will occur in the water. Each point in the water will then move clockwise in horizontal circular tracks with a radius of a few kilometres and a rotation period of about 14 hours. These rotating movements are gradually slowed down by friction in the water mass but may persist for about a week.

A satellite picture of surface water temperature in the Skagerrak showing the circulation of the surface water. The Jutland current has temperatures between 5 and 6°C, and the Baltic current, which can be seen along the south coast of Norway, has a temperature of 4.5°C. (17 March 1989, NOAA)

Currents in water can be measured in many different ways. The illustration shows an acoustic current meter which uses sound waves to determine current speeds.

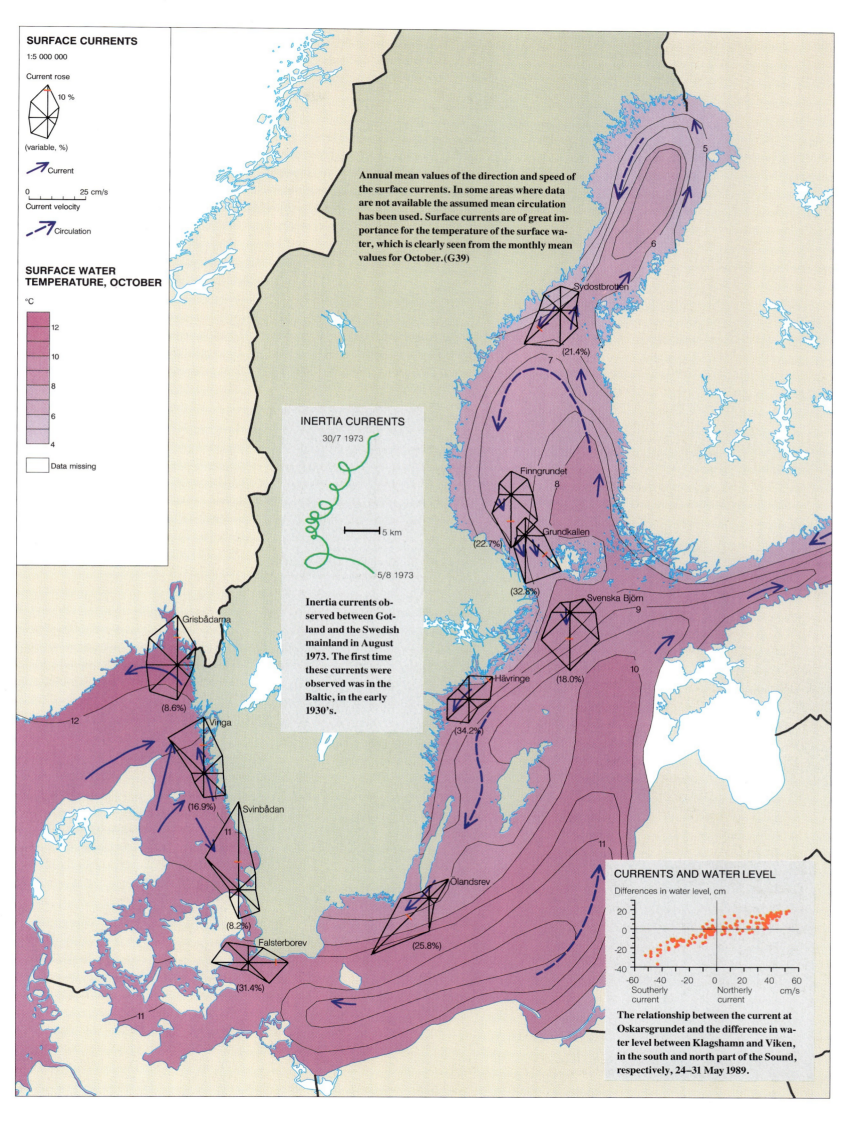

Oxygen and Hydrogen Sulphide

Oxygen is necessary for all higher organisms in the sea. When there is a lack of oxygen, then hydrogen sulphide occurs. Hydrogen sulphide is poisonous to all higher organisms and its occurrence results in dead sea beds.

The solubility of oxygen in the water depends strongly on temperature, where cold water dissolves more oxygen than warm water. As an example, it may be mentioned that saturated water at 0°C contains 70% more oxygen than saturated water at +20°C. The dependency of solubility on salinity is relatively low.

OXYGEN AND MARINE LIFE

Surface water is generally saturated with oxygen. Concentrations normally vary between 6 and 9 ml/l. The highest concentrations are generally found during the winter when the water is coldest. Oxygen enters the surface water through uptake from the atmosphere and through the photosynthesis of phytoplankton and other algae. During periods with low primary production or high oxygen consumption as a result of respiration and degradation of organic material such as dead algae, excrement, etc., the supply from the atmosphere will be of greater importance.

In deep water, where no light penetrates, photosynthesis cannot take place. In such places the oxygen is consumed by bacterial degradation of the organic material which "rains" down from the surface water. Consequently, oxygen concentrations in deep water are lower than those in surface water during the summer months when primary production is in progress. The factors decisive for oxygen conditions in deep waters are water turnover, vertical mixing, and the amount of organic material supplied from the surface water. Many species find difficulties in surviving and reproducing already when the oxygen concentrations have dropped down to about 2 ml/l.

OXYGEN CONDITIONS

In the Skagerrak and northern Kattegat there is a more or less permanent halocline which prevents a vertical mixing between surface and deep layers of water. However, this has only a marginal effect on oxygen concentrations owing to the intensive exchange of water with the North Sea and the large volume of water in relation to the organic material supplied. Nonetheless, the oxygen concentrations may locally decrease during summer and autumn in areas close to the coast where the organic load is generally larger and the water exchange poorer. Examples of such areas are some of the Bohus fiords. During the winter, in January or February, when the bottom water in the fiords is usually replaced, the oxygen levels and living conditions become improved.

The oxygen conditions in the southern Kattegat and the Baltic Proper may be compared with those prevailing in the fiords of Bohuslän. Poorer water exchange and a smaller volume of deep water, combined with a strong halocline and a heavy load cause oxygen deficiency in extensive areas. This mainly concerns deep areas of the Baltic to the east of Gotland. In these areas, the deep water, below depths of 100–130 m, is rarely replaced and may remain stagnant for 10 years or more. In 1990, the deep water to the east of Gotland had not been oxygenized for 13 years. This is the longest known period of stagnation since the late 19th century when the measurements were started, and has resulted in extensive areas with dead bottoms and hydrogen sulphide, corresponding to 10–15% of the Baltic's area (which is 3–6% of the Baltic's volume).

Long periods of stagnation start with a strong turnover of water. On such occasions the bottom water is replaced by more saline, oxygen-rich water from the Skagerrak and a strong halocline is established which prevents further water exchange for long periods. In the southern Kattegat, this being relatively shallow, the bottom water is replaced annually and the oxygen conditions are simultaneously improved. This mainly takes place during the winter through vertical mixing and through water exchange with the northern Kattegat and the Skagerrak.

The Bothnian Sea and Bothnian Bay are generally oxygen-rich as a result of a weak layering which can disintegrate during the autumn and spring and renew the deep water.

Distribution of areas of seabed with oxygen deficiency in the Kattegat, autumn, 1990. (G40)

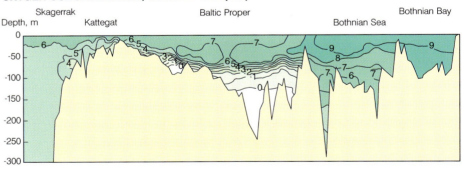

OXYGEN CONCENTRATION, SUMMER 1988 (ml/l)

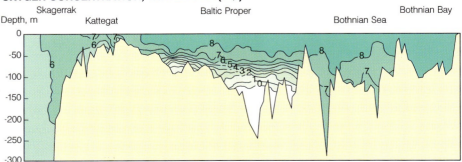

OXYGEN CONCENTRATION, WINTER 1988 (ml/l)

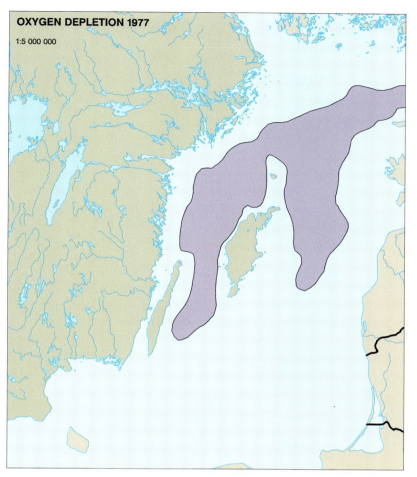
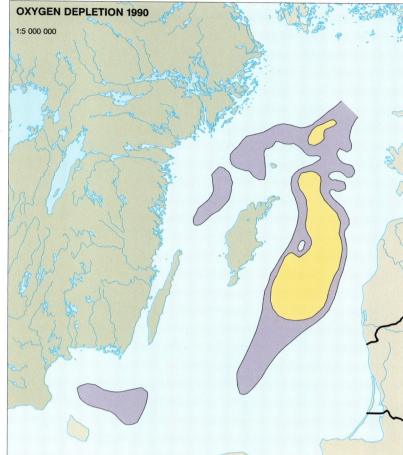

Distribution of areas of seabed with oxygen deficiency in the central parts of the Baltic, winter 1977 and winter 1990. (G41, (G42)

Oxygen (O_2) less than 2 ml/l

Hydrogen sulfide (H_2S)

CHANGES

Periods of oxygen deficiency in the Baltic are by no means a new event. Evidence has been found in the sediment which goes back as far as the 17th century and demonstrates that the bottoms have been without oxygen during certain periods. The difference today in comparison with the situation then is considered to be that the size of the extent of oxygen-free bottoms has increased during the later stagnation periods. The size of the increase is not known. However, we know that the area of the oxygen-free areas today is 5–10 times larger than the area in the early 1960's. The cause of this deterioration is a poorer water exchange and an increased supply of nutrients. The factor of greatest importance is not known with certainty.

Oxygen deficiency in the Kattegat is, on the other hand, a recent problem. During the autumn of 1980, the first indications were obtained that something was wrong; fishermen reported strongly reduced catches. In 1981, an oxygen deficiency in the southern Kattegat was observed for the first time. Earlier, a gradual decrease in oxygen concentrations in deep water during the summer had been noted but it was in 1981, for the first time, that the conditions became so poor that biological life was influenced in considerable areas. Today, oxygen deficiency is a recurrent phenomenon throughout the entire southern Kattegat.

Temperature, salinity and oxygen concentrations have been regularly recorded in the Gotland Deep.

Samples show that oxygen deficiencies occur regularly in the Baltic Sea.

Oxygen ml/l
- >3
- 3–2
- 2–1
- 1–0
- Sulphuretted hydrogen

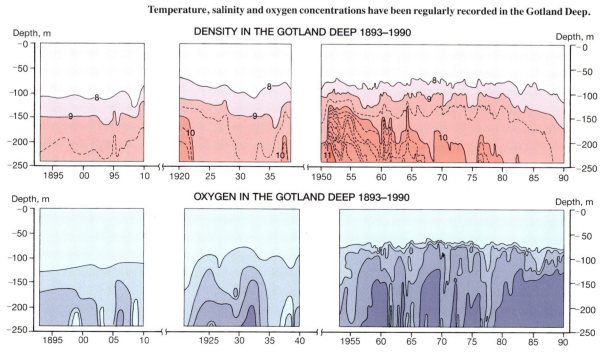

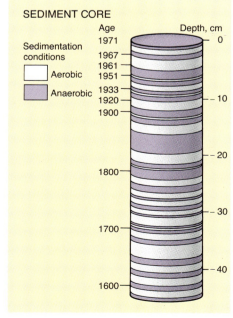

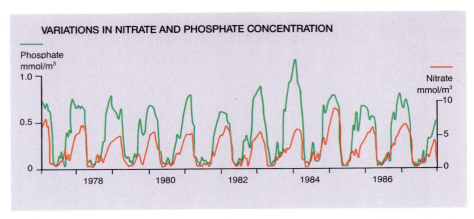

The variation of nitrate and phosphate (mmol/m³) shows a strong seasonal variation in the surface water. The measurements were made at Askö, to the south of Trosa.

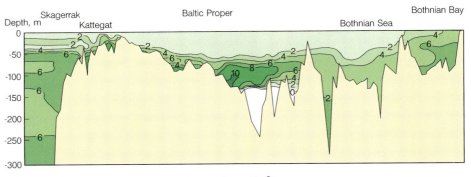

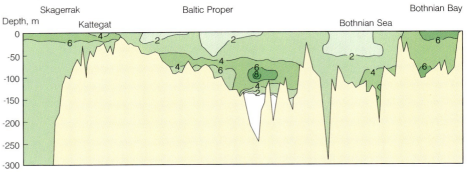

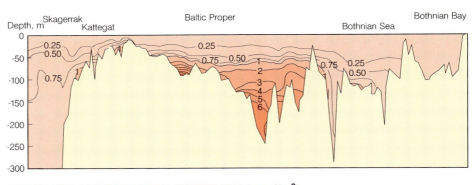

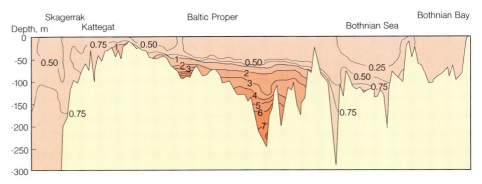

Nutrients

Organic material is largely made up of carbon, oxygen and hydrogen. Some of these elements, which are usually called nutrients, must be present in all new production of organic material. Lack of nutrients will limit growth and may cause primary production to cease completely. The two most important and most widespread nutrients are phosphorus and nitrogen. Algae and plants find it easiest to take up nutrients as inorganic compounds in the form of complex ions, nutrient salts. Nitrogen occurs both in organically bound and inorganic compounds; nitrate, nitrite and ammonium; phosphorus is only found as organic compounds and phosphate. Certain algae and bacteria are nitrogen-fixing and can utilize the nitrogen gas which is dissolved in the water. Nutrient salts enter the sea mainly with runoff from land and from the atmosphere.

Primary production starts during the spring when light conditions and stability in the water column are improved – the sea blooms. Nutrient salts are then bound in organic material, phytoplankton and plants, and concentrations in the water decrease. Most of the newly-formed organic material is generally precipitated in the surface layer in which the primary production takes place. In this way, the surface layer rapidly becomes depleted of nutrient salts and primary production decreases drastically. The organic material which remains in the surface layer is decomposed by animals and bacteria. This leads to the release of nutrient salts which can be re-used in new production. This recirculation implies that primary production never completely ceases during the summer. In some areas with an over-wintering population of zooplankton, such as in the North Sea and the Skagerrak, the recirculation is of greater importance. Growth of zooplankton in the surface water will be faster there; zooplankton consume phytoplankton and decompose them so that nutrients are released and can be re-used. In deep water, below the thermocline, there is a large reservoir of nutrient salts but the temperature stratification effectively screens them off from the surface water and thus they cannot be used in primary production. It is not until late autumn and winter, when the thermocline is broken, that the deep water becomes

INORGANIC NITROGEN AND PHOSPHORUS
1:10 000 000

Ratio nitrogen/phosphorus
- 30–
- 25–30
- 20–25
- 15–20
- 10–15
- –10

VARIATIONS IN MONTHLY MEAN
- Nitrogen (NO$_3$)
- Phosphorus (PO$_4$)

During the summer the contents of organic nitrogen and phosphate are usually low in the surface water. Only in the Bothnian Bay, where there is considered to be a deficit of phosphate, are there any particular amounts of nitrate during the summer months. The map is based on data from 1977–1989.(G43)

mixed to a greater extent and the reserves of nutrient salts in the surface can be re-filled.

NUTRIENT LIMITATION

The ratios between the basic elements normally occurring in organic material are almost constant. Roughly speaking, there are 16 nitrogen atoms to each phosphorus atom in organic material. Thus, primary production utilizes about 16 times more nitrogen (i.e., nitrate, nitrite and ammonium) than phosphorus (i.e., phosphate). In Swedish waters, the ratio between nitrogen and phosphorus (N/P ratio) usually differs from the ratio required for primary production. This implies that either phosphorus or nitrogen will be utilized first and in this way will limit primary production. The Bothnian Bay is regarded as being phosphorus limited. Here, the N/P ratio is higher than 16 and frequently reaches up to about 50, and the phosphate concentrations are relatively low (about 0.1 mmol/m^3 in the surface water during the winter). In the Baltic Proper, the Kattegat and Skagerrak, the N/P ratio is instead usually less than 16, about 10, and the phosphate concentrations much higher (about 0.7–0.9 mmol/m^3 in surface water during the winter). Primary production in these marine areas is thus considered to be nitrogen limited.

TURNOVER OF NUTRIENTS

The supply of nitrogen to marine areas is much greater than the supply of phosphorus (30–50 times). Most of the nitrogen is supplied from the air (more than 30%) whereas the phosphorus supply is completely dominated by runoff from land. Some of the nitrogen supply to the sea from the air takes place through nitrogen fixation when blue-green algae directly take up nitrogen gas dissolved in the water. In the Baltic Proper, nitrogen fixation is considered to be of particular importance, where it is responsible for 15–20% of the total nitrogen supply.

The nutrients are mainly removed from the system with the organic material which sinks down into the deep water and sediments onto the bottom, where it becomes decomposed into inorganic compounds by bottom-living animals and bacteria. It is the oxygen conditions in the bottom water which largely govern whether the nutrients are returned to the water mass or are bound in the sediment for shorter or longer periods.

Phosphorus is removed from the sea by means of the phosphate ion's ability to become bound (mineralized) in oxygenized sediment. In this way, an average of 7–10% of the sea's winter reserves of phosphorus disappear into the sediment. The variations between different marine areas are relatively small if we disregard the Kattegat and the Belt Sea/Sound, where the corresponding figure is about 30%. This depends on the large supply of phosphorus from the Skagerrak. In areas without oxygen, phosphorus is released from the sediment, and thus a net transport from the sediment to water may occur in such places.

Nitrogen is removed as a result of it becoming buried in the sediment or by denitrification. Denitrification implies that the nitrogen is released as nitrogen gas as a result of bacteria utilizing nitrate as a source of oxygen in the decomposition of organic material. This may occur in the border zone between oxygenized and oxygen-free water lying close to the bottom (the surface of the sediment). In addition, hydrogen sulphide can be formed and the sediment will become black and evil-smelling. This generally occurs in areas with limited water exchange and massive supply of nutrients, such as in the deep areas to the east of Gotland, the Laholmsbukten and enclosed fiords. Denitrification is particularly important in the Baltic Proper where about 70% of the annual supply of nitrogen is returned to the atmosphere in this way. In the Kattegat, denitrification is calculated to remove about 25–40% of the nitrogen supplied.

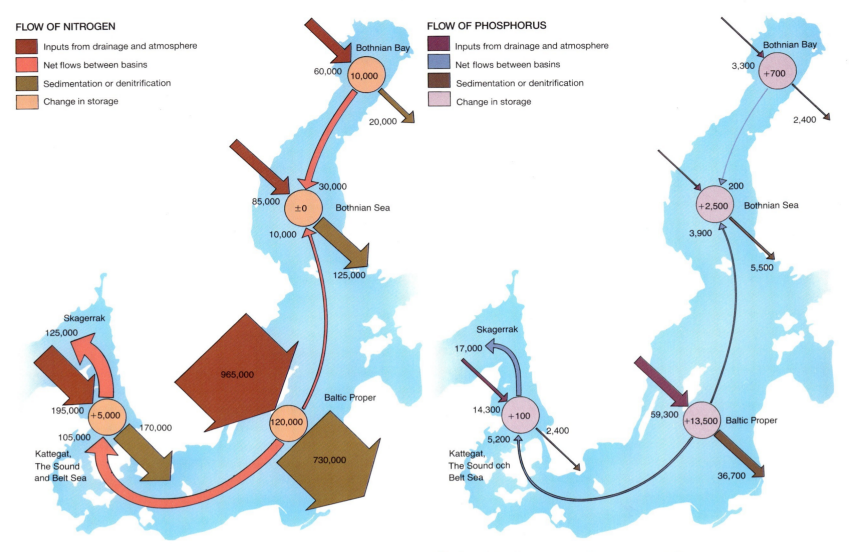

The flow of nitrogen, expressed in tonnes/year. (G44)

The flow of phosphorus, expressed in tonnes/year. (G45)

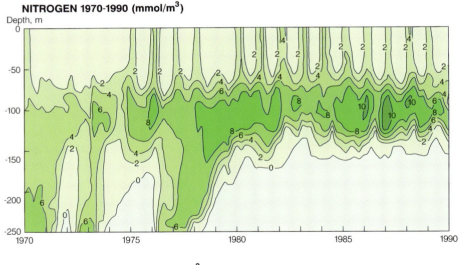

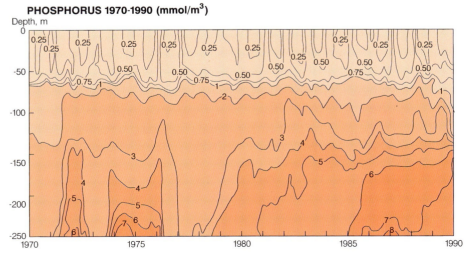

In 1976 there was a strong inflow into the Baltic of initially oxygen-rich salt water from the North Sea. The effects on the distribution of nitrate and phosphate in the East Gotland Basin can be seen.

CHANGES

The supply of phosphorus to the Baltic has increased about 8-fold during the last century, whereas the nitrogen supply has increased 4-fold during the same period. Most of this increase is due, for example, to the more intensive agriculture and road traffic in southernmost Sweden. Sewage treatment works have been built in attempts to reduce the load, but almost exclusively for phosphorus reduction. The links with the Skagerrak and the North Sea, however, reduce the importance of the changed loading in the Kattegat, where effects have become severe only locally (e.g., in Laholmsbukten). Nonetheless, the concentrations are relatively low in comparison with those found in the North Atlantic. Despite the increase of primary production in the Baltic during this period, fisheries have increased several times over and the size of the oxygen-free bottoms has increased. A continued increase in the load may result in modified species composition and succession, with major consequences for the environment.

Life in the Sea

An erosion and transport bottom at a depth of 36 m at Väderöarna, to the south of Strömstad. The clay is covered by a thin layer of sand or gravel.

Life can be found in the oceans from the surface down to the greatest depths of more than 11,000 m. Whereas chlorophyll-containing plants are restricted to living in the upper part of the sea – where light penetrates – in order to be able to photosynthesize, animals can be found at all depths. Among the animal species known on Earth today, fewer than 15% (about 160,000 species) are found in the sea. Of these, about 3,000, or less than 2%, spend their entire lives in the pelagic waters, whereas the others spend at least part of their life cycle on or in the seabed. At least 30,000 species make use of soft bottoms (sand and clay) whereas the other ca. 127,000 species live on hard bottoms made up of gravel, stones, boulders or the bare rock. Since these bottoms only make up about 1% of the surface of the seabed, it is understandable that they have a rich animal life.

Among plants, the algae have the greatest dominance. Only relatively few species of flowering plants can survive in the sea. In coastal waters, the large thallus-shaped algae (seaweed or macro-algae) are common but otherwise the unicellular forms dominate both on shallow soft bottoms and in the pelagial (as phytoplankton).

The unicellular algae are not uniformly distributed over the sea surface. Satellites equipped with instruments capable of detecting colour differences in the surface layer of the sea (colour scatterometers) provide indirect information on the algal biomass close to the surface. It is easy to see that phytoplankton abundance is high close to the continents, particularly on their west sides where nutrient-rich deep water wells up to the surface, but also in areas where cold and warm ocean currents meet, e.g., in the North Atlantic and off Japan. The central parts of the oceans are poor in phytoplankton and may almost be regarded as marine deserts.

A hard bottom in the Biotest Research Basin on the northern coast of Uppland at a depth of 5 m. Waves and currents have removed the fine material from a till bottom. The recent enclosure of the basin has created calmer conditions and deposition of organic sludge has started.

Massive algal blooms have become a regular problem in coastal areas, resulting in oxygen deficiency and dead bottoms.

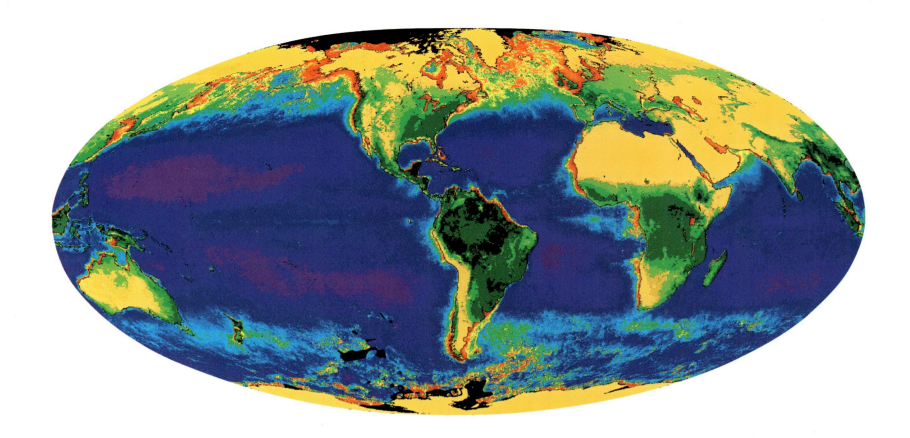

This global picture of chlorophyll concentrations, in the sea and on land, has been compiled from data obtained from more than 12,000 satellite picures, from November 1978 until June 1981. The colour scale ranges from violet/lilac (<0.1 mg/m^3) through green (ca. 0.5 mg/m^3) to red (3–30 mg/m^3).

Energy and Nutrients

Like terrestrial plants, those of the sea can utilize solar energy in converting inorganic substances, carbon from carbon dioxide (CO_2) or sometimes bicarbonate (HCO_3^-) and nutrients, e.g., nitrogen, phosphorus, silica, sulphur and iron, into organic substances required for cell development and function. In this process, photosynthesis, oxygen (O_2) is also produced and is released into the surroundings and to be utilized by the plants themselves, by animals and microorganisms in cell respiration. This is the normal process which must take place in the light part of the sea, the euphotic zone.

Since the process depends on light, this will imply seasonal variations which become greater the closer to the poles we get. In the Arctic, thus, the production period is short and perhaps only a couple of weeks long in extreme cases, whereas it continues throughout the entire year in the low latitudes.

Another limiting factor for production is the access to nutrients in the water. Normally, nitrogen is considered to be the limiting element in the open water, whereas phosphorus is limiting in the brackish coastal waters, particularly in areas such as the Bothnian Bay. However, it need not be either light or nutrients that limit growth. It may also be biological factors such as the grazing of plant-eating zooplankton. In the waters around the Antarctic, it appears that the Antarctic krill (*Euphausia superba*) plays an important role in this context. In any case, the common nutrients are not in deficiency here.

A new world of organisms was discovered in 1979 which did not base its energy requirements on sunlight but on the access to hydrogen sulphide (H_2S). Normally, this is toxic to the cells but some bacteria are able to extract energy by utilizing oxygen dissolved in the water to oxidize sulphide ions, whereupon sulphur and water are formed as final products at the same time as the energy is expended to generate organic compounds which are used by the bacteria for growth. In turn, the bacteria can live (protected) inside the cells of the animals, which can utilize the growing bacteria as food. This kind of interaction is usually called symbiosis. The cell respiration of the animal is not disturbed by the hydrogen sulphide on account of it being extremly rapidly and effectively bound to protein compounds.

This type of symbiosis takes place in several animal species found in conjunction with areas where warm (ca. 20°C) or hot (up to 350°C) water flows up through the earth's crust and forms chimney-like deposits, often called hydrothermal vents or white (warm) and black (hot) smokers. These occur as a result of deposition of salts which have been dissolved in the hot water and are subsequently precipitated when the water is cooled by coming into contact with the cold (ca. +2°C) ambient bottom water.

In areas where the plates of the earth's crust are gliding apart, and where fissures occur, the sea-water can penetrate down into the depths of the crust, become heated and then, as a result of thermal expansion, flow upwards through the seabed again.

The first hot water vents were found by geologists at a depth of about 2,000 m to the east of the Galapagos Islands in the Pacific Ocean. Later, many similar places have been found both in the Pacific and in the Atlantic.

Subsequently, biologists found that this utilization of chemical energy also occurred in shallow waters. In particular, certain mussels and bristle worms can interact with bacteria in the same way in order to make use of energy in areas where hydrogen sulphide can be formed as a result of oxygen deficiency in the bottom water or in the upper layers of the bottom sediment as a result of poor water turnover. This also applies, for example, to the species of one of the mussel genera commonly found in Swedish west coast waters, *Thyasira*.

It has also been discovered that other bacteria can utilize methane (CH_4) as an energy base. Methane is usually formed during bacterial degradation of organic material in bottom sediment and such sources of methane have been found in many places. The first discovery was made in deep water in the Gulf of Mexico, but subsequently methane seeps have also been found in the North Sea and in the Kattegat, among other places. Once again, a specially adapted animal world develops.

Some researchers consider that life itself might have been created in conjunction with the hot water vents, and if that is the case, then life has been created more than once during the history of the globe.

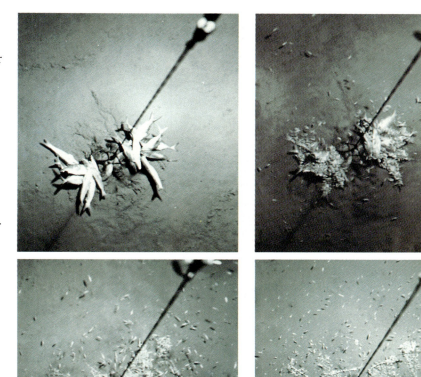

A series of photographs, taken at a depth of about 10,000 m in the Philippine Trench, shows how baits of fish can be stripped to a skeleton within a period of slightly more than 16 hours by amphipods. The photos were taken 10 minutes, and 4, 8, and 16 hours after the baits reached the bottom.

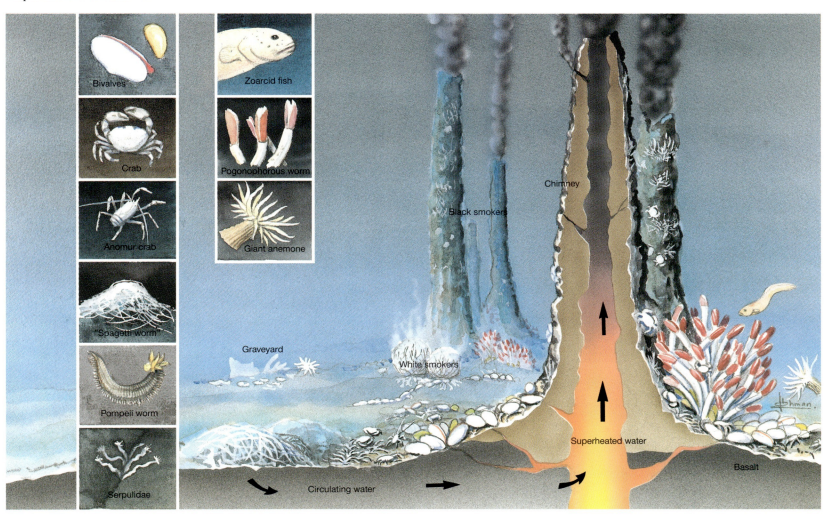

Illustration of a deep sea community in which the energy supply is based on utilising sulphur bacteria capable of oxidising hydrogen sulphide. The water passes through chimney-like vents containing hydrogen sulphide and has a temperature of more than 300°C.

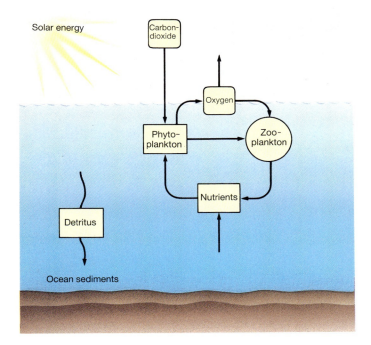

The Biological Pump

The biological pump implies the role of marine organisms in the circulation of certain elements in the sea and by means of the interfaces between the sea and the atmosphere, and between the sea and the sediment on the sea floor. This is particularly interesting in relation to the "greenhouse" gases (e.g., carbon dioxide, methane, freons) and the role of the sea for the carbon dioxide cycle is now the subject of particular study. It is probable that marine organisms are just as important as purely physical mixing and dissolving of these gases in sea-water in helping to reduce the surplus of carbon dioxide produced by man in, for example, combustion of fossil fuels. The route taken by the carbon through photosynthesis and biological circulation to carbon dioxide, water and different organic degradation products is remarkably effective. Some of this material is recirculated in the upper layers of the sea whereas another part is brought down to greater depths to be either deposited on the bottoms or slowly returned to the surface layers.

It is estimated that about half of the surplus of 5–6 billion tonnes of carbon (in the form of carbon dioxide) released annually through anthropogenic activities to the atmosphere is absorbed by the sea. If this did not take place, then the carbon dioxide content in the air would increase much faster than is the case today, probably resulting in increased global heating. Thus, the biological pump of the sea helps to moderate our climate.

Ecosystems and Ecosystem Models

The science of the mutual relationships of organisms, including the relationship with the ambient physical-chemical-geological environment, is called ecology. Organisms depend upon each other, since some are producers capable of converting light energy or chemical energy into organic material, whereas others are consumers. These obtain their nutrition by eating other organisms or degradation products of such organisms. This situation is frequently described as a nutrient pyramid, with producers (usually phytoplankton or bottom-living algae) making up the wide base, above which we find zooplankton which live on producers, then the animals living on zooplankton, and finally the top or large predators at the narrow peak of the pyramid. As a general rule, about 90% of the energy is lost between each level in the pyramid. This applies both in the sea and on land.

The relationships between organisms are frequently complex and drawings can be made of them as models in food chains, in complicated food webs, or in simpler ecosystem models. In these models, other factors may also be included, e.g., energy sources, nutrients, water movements in the form of up-welling and down-welling, migration of fish, and different types of sea floors.

The important role of bacteria and other very small organisms in the energy turnover in the water mass and on the bottoms has only been realized during the past decade.

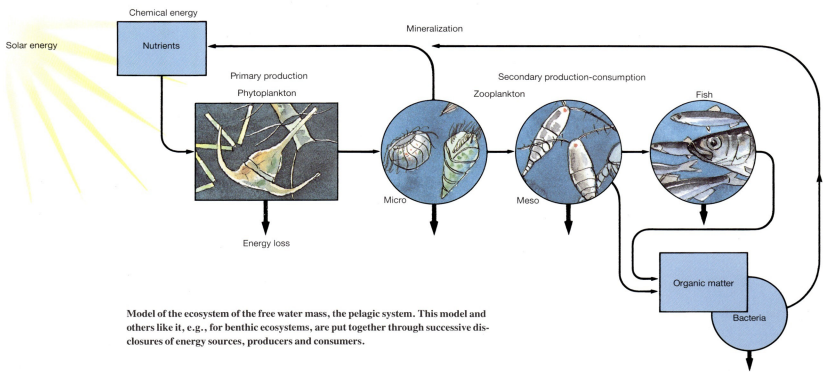

Model of the ecosystem of the free water mass, the pelagic system. This model and others like it, e.g., for benthic ecosystems, are put together through successive disclosures of energy sources, producers and consumers.

ECOSYSTEM, PRIMARY AND SECONDARY PRODUCTION

 Supply from river run off and direct industrial discharges

 Pelagic primary production

 Benthic primary production

 Bacteria

 Zooplankton

 Fish

 Benthos

 Sedimentation

The energy resources of the ecosystem (expressed in grams carbon/m² and year) consist of the sum of primary production (i.e., production of algae and phytoplankton) together with the supply of organic matter through runoff and industrial emissions. This energy is used in secondary production, either directly by, e.g., zooplankton which eat phytoplankton, and indirectly by fishes eating the zooplankton.(G46)

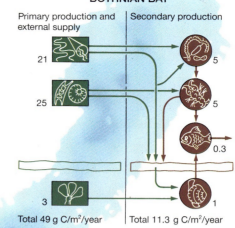

BOTHNIAN BAY — Total 49 g C/m²/year | Total 11.3 g C/m²/year

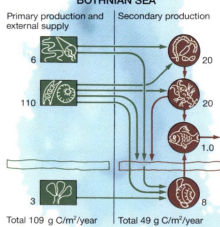

BOTHNIAN SEA — Total 109 g C/m²/year | Total 49 g C/m²/year

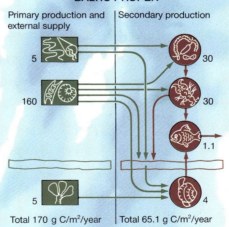

BALTIC PROPER — Total 170 g C/m²/year | Total 65.1 g C/m²/year

KATTEGAT — Total 237 g C/m²/year | Total 143.3 g C/m²/year

77

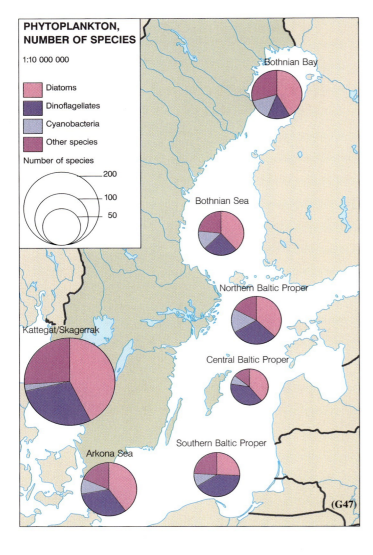

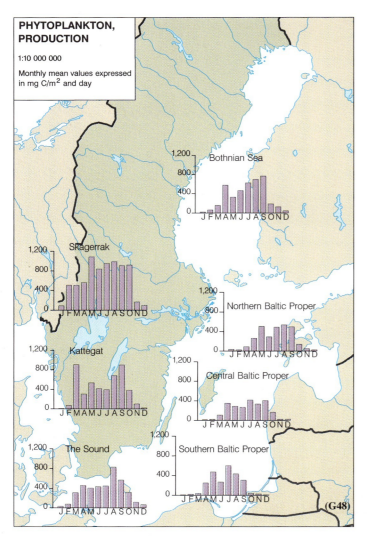

Phytoplankton

Phytoplankton are the basis of life in the sea. They convert solar energy into the organic material which is the fuel for other organisms in the sea. They are unicellular algae varying in size from some thousandths of a millimeter up to about 2 mm. Phytoplankton have an extremely short generation time – ranging from half a day to about a week. In situations of rapid growth large populations develop in a short time. This is called plankton blooming.

There is a vast abundance of plankton algae species, with many thousands of species divided into a large number of groups. In the seas around Sweden we can find species representing almost all groups.

Diatoms have a cell wall of silica which is shaped like a small tablet-box with a bottom and a lid. The silica shell is richly patterned. Cells are frequently linked into chains.

Dinoflagellates have two flagellas which allow them to move. About 25 of the ca. 1,000 dinoflagellate species are capable of forming toxins which may affect other marine organisms and, secondarily, human-beings.

Cyanobacteria (blue-green algae) are closely related to bacteria and have their major distribution in fresh water and tropical seas. In addition, they are very common in the Baltic, where several species form huge blooms regularly. Some species are capable of utilizing the nitrogen gas in the atmosphere and are therefore independent of the nitrogen compounds present in or supplied to the sea.

DISTRIBUTION, BIOMASS AND PRODUCTION

The species composition of phytoplankton is influenced by the extremely large changes in salinity found when moving round the coasts of Sweden, from the innermost parts of the Bothnian Bay out to the Skagerrak. In the Bothnian Sea and the Baltic Proper there are many species which are the same as, or similar to, those in fresh waters. Going north along the west coast from the Sound, or even from the area to the south of Skåne – the Arcona Basin – it is obvious how marine species successively increase in importance. In the northern Kattegat and in the Skagerrak there are only small remnants of the Baltic forms. The largest change in salinity takes place in the Sound, and it is also here that the largest changes in species composition of phytoplankton are found.

Certain species of the Baltic forms are more strictly bound to the Baltic. The diatom *Navicula vanhoeffenii* and the colony-forming dinoflagellate *Peridinella catenata*, for example, are only rarely found to the west of Bornholm. Two of the characteristic species of the Baltic, the cyanobacteria *Aphanizomenon flosaquae* and *Nodularia spumigena*, withstand a slightly higher salinity. They may occur in large blooms in the western Baltic and even in the Kattegat. For most Baltic species it is difficult to draw definite boundaries for their distribution since they are transported out of the Baltic with the surface current. It is also difficult to define distinct limits for the Kattegat/Skagerrak species since they are transported by inflowing salt water and forced down to greater depths in the Sound. Frequently, they are taken further in with the salt deep water and are found far up in the Baltic, even though their density decreases strongly once the Sound has been passed.

The species composition of phytoplankton changes not only geographically but also during the year. This temporal change in species composition and in the amounts of the different plankton algae is called succession. Among the dominating algal groups in the seas around Sweden we find a succession where diatoms dominate at the start and end of the growing season. After the vernal diatom bloom, the importance of dinoflagellates and other small flagellates increases. In addition, there are cyanobacteria in the Baltic during the summer. In the Skagerrak/Kattegat the dinoflagellates reach a maximum in September when they may occur in such large numbers that the water is coloured brown-red. It is during these dinoflagellate blooms that phosphorescence can sometimes be seen.

The periods when the algae occur reflect their environmental requirements. Diatoms dominate, relatively speaking, in cold and nutrient-rich waters and dinoflagellates in warm and nutrient-poor waters. The dominating cyanobacteria in the Baltic reach their maximum in warm water with a low concentration of inorganic nitrogen. Since they can fix nitrogen from the air, they are able to outcompete many other species.

In the seas around Sweden, the

A representative of the diatom genus *Coscinodiscus*.

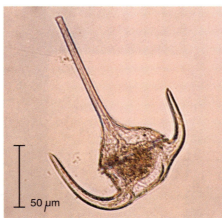

The dinoflagellate *Ceratium tripos*.

PLANKTON

Plankton are the organisms, both plants and animals, which passively drift with the movements of the water. Although some of them are capable of swimming, they cannot swim well enough to counteract the movements of the water.

Plankton are not uniformly spread throughout the sea. They occur in "clouds", the sizes of which vary depending on the effect of biological and physical factors. The ecological interaction with, e.g., the grazing of zooplankton on phytoplankton, may be decisive for the size and extent of a "cloud of phytoplankton". Insolation or light, temperature and salinity in the sea are physical factors which determine where and when different species of plankton occur.

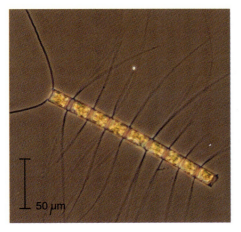

Chain of the diatom genus *Chaetoceros*.

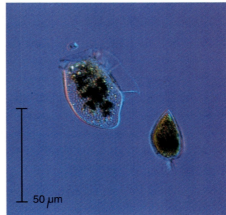

The dinoflagellates *Prorocentrum micans* (left) and *Dinophysis acuta* (right).

HOW DOES THE SEA BLOOM?

The phenomenon whereby plankton algae grow rapidly and in a short period expand into vast populations is called algal bloom. Certain conditions must be fulfilled for explosive growth to take place. Strong sunshine and calm weather are important. In addition, the algae must have a supply of nutrients such as nitrogen and phosphorus. This may take place as a result of anthropogenic excess fertilization but may also be achieved as a result of the naturally nutrient-rich deep water is brought to the surface where the algae develop.

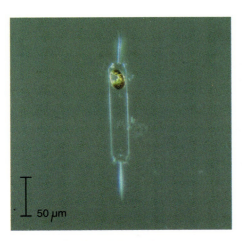

The diatom *Ditylum brightwellii*.

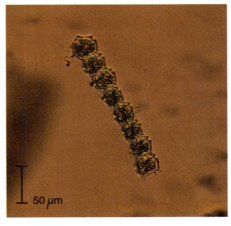

Chain of the dinoflagellate *Peridinella catenata*.

MARINE EUTROPHICATION

This involves excessive inputs of fertilizing elements entering the seas, which causes algal production to increase. Eutrophication normally has negative effects, e.g., oxygen deficiency, when the large quantity of algae are to be decomposed. This, in turn, may lead to fish and bottom-living animals suffering. Moderate eutrophication may have positive effects as the animals in trophic levels above plankton algae may make use of the increased algal production.

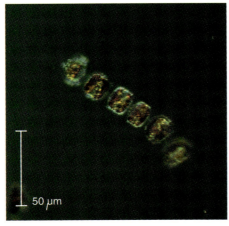

Chain of the diatom genus *Thalassiosira*.

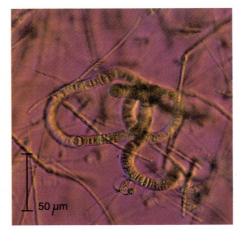

Chain of the cyanobacteria *Nodularia spumigena*.

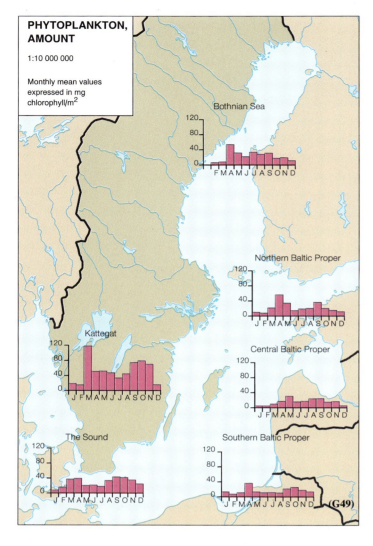

PHYTOPLANKTON, AMOUNT
1:10 000 000

Monthly mean values expressed in mg chlorophyll/m^2

RED TIDE

This refers to extremely large accumulations of plankton algae which give the water a reddish colour. The high concentration of algae occurs either as a result of strong growth or of algae accumulating as a result of currents and winds.

Red tide in the Skagerrak, June 1990.

changes in salinity are a dominating factor and are decisive for the number of species. In waters of high salinity, as well as in fresh water, the number of species is high. Species of marine origin and freshwater species enter the brackish waters of the Baltic from different directions, but many of them cannot survive the osmotic stress caused by higher or lower salinity. Consequently, the number of species of phytoplankton is relatively low in large parts of the Baltic Sea.

In the same way as the number of species increases out towards areas with higher salinity, an increase also takes place in phytoplankton production. In the Bothnian Sea, annual production is only one-third of that in the Skagerrak. The amount of phytoplankton (biomass), on the other hand, does not differ to the same extent between different marine areas.

RECENT CHANGES

The seas around Sweden have become increasingly loaded with nutrients during the 20th century. This excessive fertilization, eutrophication, has its primary effect on phytoplankton which utilize nitrogen and phosphorus for increased production. Since phytoplankton production has been investigated only during recent decades, it is difficult to give a definite assessment of the degree to which eutrophication has resulted in increased algal production. In addition, there is also the natural variation from one year to another which may be so large that a change caused by eutrophication is masked. Despite this, there are both direct and indirect indications that changes have taken place in the marine phytoplankton.

Along the Swedish west coast, it has been found that phytoplankton production has probably increased by about 10% during the last 50 years. An increase of the same size has probably also taken place in the Baltic, but it has not been possible to demonstrate this numerically. The clearest indication of changes is found in the increasing number of algal blooms. In the Baltic, cyanobacteria (blue-green algae) have always bloomed during the summer but they seem more common today than earlier. They also appear to be toxic more frequently. The west coast rarely had comprehensive algal blooms earlier, but today they occur more or less regularly. Usually dinoflagellates form these blooms and toxic substances are observed every year.

A clear sign that recent changes have taken place among the phytoplankton is that during the 1980's several species were found which have not earlier been reported from Swedish waters. Several of them are dinoflagellates known from other parts of the world. Some of them are unable to survive in the low salinity of the Baltic Sea and are therefore only found along the west coast. Others have greater tolerance and are dispersing slowly into the Baltic.

One of the most noticeable changes found in phytoplankton was the comprehensive *Chrysochromulina polylepis* bloom in large parts of the Skagerrak/Kattegat during the spring and early summer of 1988. This bloom, of an organism known in the area but which had not earlier caused problems, resulted in the poisoning of other marine organisms such as algae, mussels and fish.

PHOSPHORESCENCE

Some dinoflagellates are capable of emitting flashes of light which are produced by means of a chemical reaction in the cell. Normally, a mechanical impulse is required to start the chemical reaction. The flashes of light are extremely weak and thousands of dinoflagellates must be disturbed simultaneously for the human eye to register the light. Phosphorescence may be seen in the wake of a boat on a late-summer evening at the west coast.

Zooplankton

Zooplankton include both unicellular and multicellular animals. The unicellular animals are often only a few hundredths of a millimetre in size whereas the multicellular, which also include jellyfish, may have a diameter of more than one metre. The zooplankton which remain in the free water mass, the pelagial, throughout their lives, are called holoplankton. Many bottom-living animals have larval forms which, during a limited period of time, live as zooplankton in the pelagial. These zooplankton are called meroplankton. Fish larvae and fish eggs are also included among the zooplankton and have the common name of ichthyoplankton.

The classical picture describes the flow of energy in the pelagial as a food chain where phytoplankton are eaten by zooplankton which in turn are eaten by fish, etc. Recent discoveries in relation to the energy flow through bacteria and micro-forms of phytoplankton and zooplankton have resulted in the energy flow in the pelagial today being described as a richly branched nutrient web, the pattern of which changes during the year depending on species composition, temperature, food availability, etc.

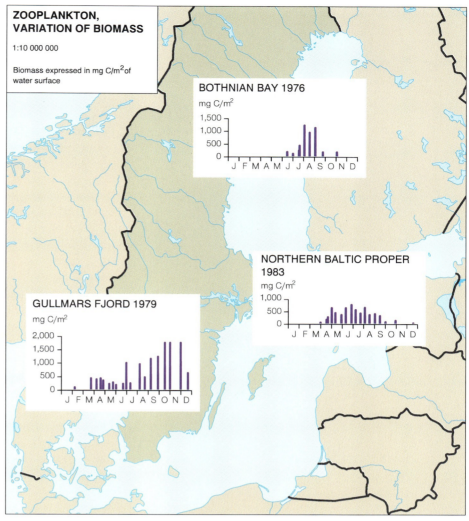

The biomass figures apply to zooplankton larger than 0.09 mm, with the exception of scyphozoans and ctenophores. (G50)

Zooplankton in the Skagerrak and Kattegat, including copepods, cladocerans, chaetognats and planktonic larvae of benthic fauna.

81

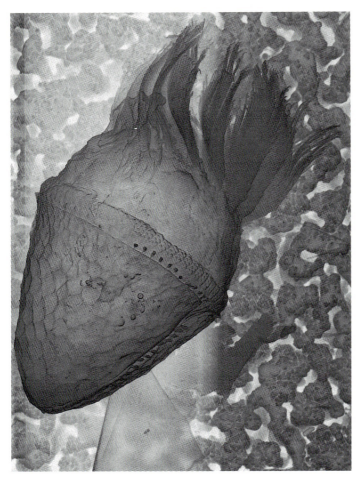

Electron microscopic picture of a ciliate commonly found at the Swedish west coast, *Strombidium reticulatum*.

Egg-carrying rotifer (*Synchaeta* family). During the late winter the rotifers can periodically completely dominate the zooplankton community.

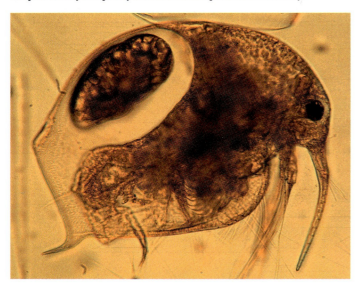

Cladoceran belonging to the *Bosmina* family. This family is very abundant in the surface water of the Baltic during the late summer.

DISTRIBUTION

In the seas around Sweden it is mainly the salinity which limits the distribution of different species. On the west coast, there is a mixture of marine and Baltic species which enter the Kattegat through the Sound transported by the Baltic Current. A few purely marine species live in the Baltic together with brackish water species. In addition, there are a number of fresh water species which also can survive in brackish water. As we go north in the Baltic the marine species decrease and the fresh water species increase.

The species composition in a certain area is, however, rarely constant but is influenced by the hydrographic conditions prevailing at that time.

SEASONAL VARIATIONS

During the winter, when there is low availability of food, there are few zooplankton in the pelagial. They over-winter either as resting eggs which lie on the bottom or as a result of older individuals migrating down to the deep waters and entering some kind of hibernation. In order to survive they utilize the energy reserves of fat which they have stored earlier during the autumn. In connection with, and immediately after, the spring bloom of the phytoplankton, there is an increase in the number of zooplankton in the pelagial. The number of species is largest during the summer when the density of zooplankton is large also in deeper layers of water. During the autumn, there is a successive reduction of both individuals and number of species, and the zooplankton prepare for the winter.

BIOMASS AND PRODUCTION

Methods of directly measuring zooplankton biomass or production in the sea have not yet been devised but estimations can be made on the basis of assumptions of the volume and growth of the organisms. From such estimations we can see that there is an annually recurrent pattern in the seasonal variation of, e.g., biomass, and that this pattern is different in the Skagerrak/Kattegat, the Baltic Proper and the Bothnian Bay.

CILIATES

Ciliates are unicellular zooplankton. They comprise two groups, naked ciliates and tube-living tintinnids. The naked ciliates are between 0.01 and 0.05 mm large. The tintinnids are of the same size but their tubes may be as long as 0.4 mm. They live on bacteria and micro-forms of phytoplankton (flagellates) and are found in the upper layers of the water column along the entire Swedish coast. They multiply by division and can, under favourable conditions, divide several times per day.

ROTIFERS

The rotifers are small multicellular zooplankton, most of them between 0.07 and 0.5 mm long. They are more abundant in the upper layers of coastal water than in the open sea. The most commonly found families are *Synchaeta* and *Keratella*. Rotifers over-winter in the form of resting eggs which lie on the bottom. From these eggs females are hatched which swim up to the surface. In turn, they lay new eggs which also produce females when they hatch (virgin birth or parthenogenesis). Under favourable conditions the generation time from egg to new egg is only 2–3 days. In addition, several species can grow rapidly even at low temperatures. Towards the end of the season, males also develop, and they fertilize the females. These then produce new resting eggs which sink to the bottom where they over-winter.

The rapid growth of a rotifer population enables it to respond rapidly to, e.g., increased food availability. In the spring, when there are still fairly few other zooplankton in the pelagial, the rotifers may sometimes completely dominate the zooplankton community.

CLADOCERANS

The cladocerans have the same type of life cycle as the rotifers with over-wintering of resting eggs and a largely parthenogenetic reproduction. Cladocerans also occur mainly in surface layers but prefer warmer water and are therefore mainly found there during the summer. Periodically they may be extremely numerous.

The most abundant species differ from area to area; *Daphnia* and *Bosmina* are most common in the Bothnian Bay, *Bosmina*, *Evadne* and *Podon* in the Baltic and *Evadne* and *Podon* in the Skagerrak/Kattegat. They are all between 0.5 and 1 mm long.

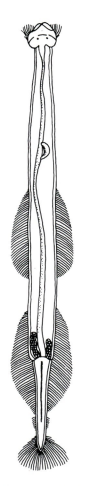

A chaetognat – *Sagitta elegans* – commonly found in both the Baltic Sea and on the west coast of Sweden.

COPEPODS

Copepods constitute the dominating group of zooplankton. They occur along the entire Swedish coast and are also common in the open sea. Apart from the groups described above, they can be found throughout the water column from the surface to the seabed. The life cycle of the copepod covers 6 larval stages (nauplia) after which they change dramatically (metamorphosis) and after a further five moults, they become adult. They multiply only sexually and the generation period from egg to new egg is about two weeks at the shortest.

Older individuals of many copepod species migrate daily through the water column. During the day, they remain in deep water where darkness protects them against predators. At dusk they swim towards the upper layers to feed. Most species live on phytoplankton or on both phytoplankton and smaller zooplankton (e.g., ciliates and rotifers). In Swedish waters there are few species which are exclusively carnivorous.

Some copepods over-winter by means of resting eggs whereas others over-winter as a result of older individuals migrating in the autumn towards deeper water where they enter a condition similar to hibernation.

The smallest nauplius stage of copepods is about 0.07 mm long. Adult copepods, depending on species, are sized between 1 and 4 mm.

CHAETOGNATS

Chaetognats are commonly found along the Swedish west coast and in the southern Baltic. They are sparsely found in the northern Baltic where the low salinity restricts their distribution. Chaetognats are predators and the two most commonly found species in Swedish waters are *Sagitta elegans* and *S. setosa*. They reach a maximum length of 40 and 50 mm, respectively, and both have a life-span of one year.

CTENOPHORES

Ctenophores are easily recognized by the eight rows of combs made up of ciliary plates which extend from pole to pole along the jelly-like body. Ctenophores are not jelly-fish in the true meaning, and it is probably their consistency which has led to this misapprehension. The most characteristic ctenophore in Swedish waters is *Pleurobrachia pileus* which, both in size and form, is similar to a transparent gooseberry. During autumn and winter, this species is found in large numbers on the Swedish west coast, but is also found in deep waters of the Baltic. The two species *Beroe cucumis* and *Bolinopsis infundibulum*, both of which are much larger than *Pleurobrachia*, are also commonly found along the west coast even though they are not as abundant as *Pleurobrachia*.

SCYPHOZOANS

The true jelly-fish, the scyphozoans, are represented in Swedish waters by *Aurelia aurita*, *Cyanea lamarckii* and *Cyanea capillata*. *Aurelia aurita* is commonly found both along the west coast and in the Baltic. *Cyanea lamarckii* is not found in the Baltic and *Cyanea capillata* is found there only sporadically. Both *Cyanea lamarckii* and *Cyanea capillata* are, however, common in the Kattegat/Skagerrak.

The youngest individuals of *Aurelia aurita* are found already during late autumn but it is not until spring that true growth starts. *Aurelia aurita* is subsequently commonly found during the entire summer. *Cyanea lamarckii* becomes common on the west coast during May and June and *Cyanea capillata* is found during the whole summer until late autumn. The larvae of jelly-fish develop during summer and the early part of autumn. For a short period they swim freely as plankton but soon attach to a suitable surface such as rock, stones, mussel shells, etc. There they soon develop into a polyp which, during late autumn (for *Aurelia aurita*) or early spring (*Cyanea*) give rise to the next generation of jelly-fish.

TUNICATES

The most commonly occurring planktonic tunicates are the appendicularia. They are characterized by a clear or milky-white "house" consisting of a finely-meshed net through which the food is filtered-off, and also by their "tail". This "tail" is in fact a backbone similar to those in vertebrates.

Two species of appendicularia are common in Swedish waters; *Oikopleura dioica* and *Fritillaria borealis*. The length of the "house" in *O. dioica* is 0.5–1 mm and the tail is 2–4 mm. This species is found along the entire west coast. In the Baltic, the distribution of the species is restricted to the southernmost parts where it enters with the inflow of salt water from the Kattegat. In *F. borealis*, the house is about 1 mm long and the tail is short and wide in comparison with that of *O. dioica*. This species is common along the west coast and in the Baltic up to the southern part of the Bothnian Bay.

A copepod belonging to the *Calanus* family. These copepods are commonly found on the Swedish west coast.

The ctenophore – *Pleurobrachia pileus* – in its natural environment. The rows of combs, with which the ctenophore propels itself, and the tentacles are clearly visible.

Vegetation

Most of the plants in the sea are algae. Altogether there are almost 15,000 plants in the seas of the world, of which about 95% are algae. Algae are the oldest plants in evolution, but at the same time they are the least known to the general public.

In contrast to flowering plants, algae lack flowers, seeds and roots, among other things. They reproduce with unicellular spores, or eggs and nutrients are absorbed directly from the water through their entire thallus surface. Similar to flowering plants, all algae contain chlorophyll a. With the help of chlorophyll and with energy from sunlight they build up organic substances from carbon dioxide (photosynthesis). The depths at which algae are found in the sea are therefore governed by the depth reached by sunlight. Along the Swedish coasts algae can grow maximally down to about 25–30 m deep in clear water off the coast but only down to a couple of metres in turbid water such as is found in the proximity of river estuaries and outlets.

In addition to carbon, algae require, like all plants, nutrients such as nitrogen and phosphorus. Both concentrations of, and the balance between these elements are important for how much and what grows. Growth and uptake rates of nutrients are faster in small algae which have a relatively large surface in relation to their volume. Thus, for example, filamentous algae, which have a large active surface, are the ones which dominate in areas with high concentrations of nutrients and this also contributes to differences in algal vegetation found between inner and outer archipelagos, both in the Kattegat/Skagerrak and in the Baltic Sea. The same also applies if one and the same area is exposed to increased eutrophication.

Macroalgae (algae which can be seen with the naked eye) are divided into three main groups: Green, brown and red algae, according to their dominating colours. In green algae, chlorophyll gives them their green colour, whereas chlorophyll in the other two groups, as in most microalgae, is hidden by other pigments. Their function is to absorb light energy of other wave lengths and to transfer energy to the chlorophyll. There are also major differences in reproductive strategies and in stored compounds. Consequently, differences between the diverse algal groups are therefore much larger than between terrestrial plants, such as between a buttercup and pine-tree.

DISTRIBUTION OF MACROALGAE

Almost all organisms in the Kattegat/Skagerrak and most of those in the Baltic Sea are of marine origin, but a smaller number of freshwater species have migrated out into the Baltic Sea, mainly into the Bothnian Bay, during the varying geological history of this area. Since the Baltic is a young sea geologically, only a very small number of unique (endemic) species have had time to develop. The marine algae survive the changes in salinity to different extents, and consequently the number of species decreases the closer to the Bothnian Bay we get, and many important algal groups are completely absent in the Baltic Sea–Bothnian Bay area. Not only does the Baltic lack lobsters, crabs, starfish and sea urchins, but also most of the large brown algae are absent in almost the entire Baltic Sea, and neither are there any calcareous red algae. Generally, the red algae as a group are sensitive to low salinity which means that their proportion of the total decreases more than that of the other groups. The algal vegetation at the extremity of the Bothnian Bay consists almost entirely of green algae, where most species are in fact of freshwater origin. There are also many flowering plants on the sediments – the more brackish the water the larger the number of species present.

Many marine species are influenced by the low salinity before they disappear completely. Several of the red algae, for example, become sterile, grow as loose-lying plants in the brackish water of the Baltic Sea, and multiply only by means of fragmented parts of the thallus growing out into new plants. Many, but not all, algae are also smaller in brackish water since they use large amounts of energy to survive in the low salinity – the same phenomenon as can be seen among mussels.

LOCAL DIFFERENCES AND ZONATION

The Atlantic coast is characterized by a pronounced tidal water zone which commonly has an amplitude of some metres, exceptionally reaching 15 m. Along the Swedish coast the differences in tidal water are only about 30 cm at the coast of the province of Bohuslän and less than 5 cm in most of the Baltic Sea. Consequently, a number of the typical Atlantic tidal water species is not found along the Swedish west coast.

The uppermost limit of the sea shore proper in the Kattegat/Skagerrak is usually marked by a white belt of rock barnacles, a sessile crustacean. Species growing in the tidal water zone on the Swedish west coast depend largely on how wave-exposed the place is. Large brown algae such as bladder wrack, knotted wrack and flat wrack (*Fucus vesiculosus*, *Ascophyllum nodosum*, *Fucus spiralis*) form belts on more or less protected shores and are accompanied downwards by a belt of serrated wrack (*Fucus serratus*). On these perennial seaweeds, one often finds short-lived filamentous brown and red algae (epiphytes), and in more nutrient-rich water also green algae. Hydroids and small coiled serpuloid tube worms are other epiphytes on the wracks. In places strongly exposed to waves, on the other hand, these seaweeds are lacking and the shore is dominated by short-lived red algae together with the common mussel (*Mytilus*). From a few metres deep, the rocks are covered by belts of the large kelp species (*Laminaria spp.*) with their leaf-like blades (lamina), often more than one metre long, which in two of the species are sliced at the top. As on land, there are plants in several layers, which can easily be seen when the thalli of the large algae are pushed to one side and the under-vegetation of shade-tolerant species becomes visible. The lowermost layer grows directly on the rock and consists of calcium encrusted red algae which look as if they have been painted onto the rock surface. From 5–10 m deep, most of the algal vegetation consists of red algae, and here there are more than 100 different species in a diversity of forms ranging from thin leaf-like to cartilagenous or extremely finely-branched algae. Many of them grow as epiphytes on other algae. The variation of macroalgae in these northern waters is even greater than that of coral reefs. The deeper we get the more the sessile animals start to dominate, e.g., sponges and seasquirts take over the space on the rocks. The macroalgae growing in the deepest parts are a small number of crustose red algae.

ALGAE, NUMBER OF SPECIES
1:20 000 000

- Red algae
- Brown algae
- Green algae

Number of species
100
50
0

N Bothnian Bay
S Bothnian Bay
Uppland
Södermanland
Bohuslän
Halland
Blekinge
The Sound

Number of red, brown and green algal species larger than 1 cm. (G51)

84

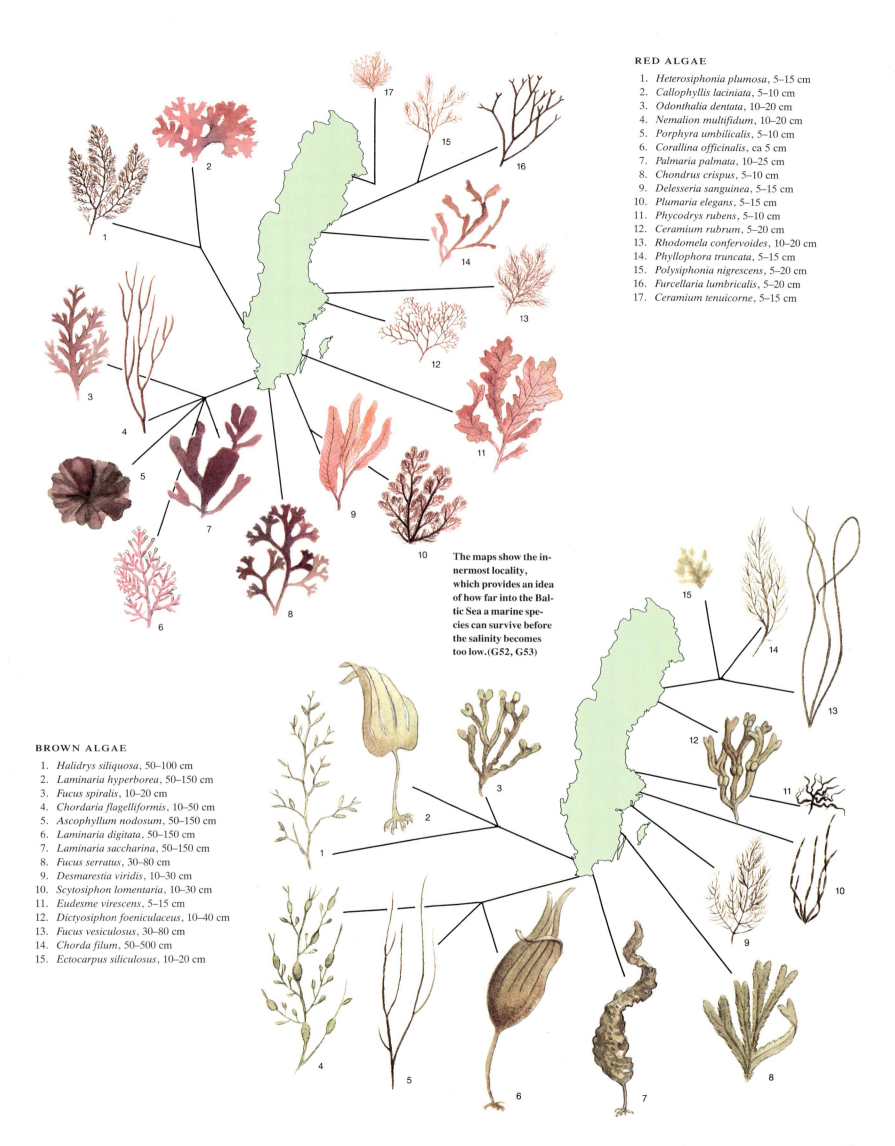

Japweed
(*Sargassum muticum*)

Eel grass
(*Zostera marina*)

Despite the lack of tidal water on the Swedish east coast, there are considerable alterations in water level there and differences of up to 1–2 m may be registered during the year depending on low or high pressure conditions. Since the shore may emerge for a period of days or weeks, and in addition is frequently exposed to ice erosion during the winter, there are hardly any perennial species on the uppermost part of the shores in the Baltic Sea. Instead, in the Baltic this zone is covered by different filamentous algae during the seasons, with a dominance of the brown alga *Pilayella littoralis* in the spring, the green alga *Cladophora glomerata* in the summer and the red alga *Ceramium tenuicorne* in the autumn. The perennial bladder wrack is found in the Baltic only at depths below ca. 0.5 m. In comparison with the west coast, the deeper algal vegetation on these coasts is less rich in species and is dominated by a small number of decimeter-large, more or less cartilagenous, red algae such as *Furcellaria lumbricalis* and *Phyllophora spp.* together with a small number of finely-branched species, and deepest of all we can frequently find the brown alga *Sphacelaria arctica*. Instead of the calcareous crustose red algae, the stones are covered by crusts of brown algae.

Nearly all the large seaweeds are missing in the northernmost part, the Bothnian Bay, and the filamentous green algae cover the rocks. Also the water moss *Fontinalis* is an important species in this region. Some algae are found deeper in the Baltic Sea than on the Swedish west coast.

About ten macroalgal species new to Sweden have immigrated to the west coast during the 20th century and are presently spreading.

PLANTS ON SAND AND MUD BOTTOMS

The vegetation seen on shallow sand and mud bottoms is usually dominated by flowering plants, in the Baltic Sea also by stoneworts (charophytes), a special group of green algae which are attached in the substrate with their root-like formations. Other macroalgae are attached to stones or to shells of mussels or snails, or can be found as unattached individuals which may continue to live and which sometimes form large carpets on the bottoms of protected coves.

In the Skagerrak/Kattegat, dense meadows of the flowering plant eel grass (*Zostera marina*) are formed. With their rich abundance of food and places to hide in, these meadows are of major importance for small animals and fish fry. Where gravel/pebbles or shells occur on the sediment, it is also possible to find the long Sea lace (*Chorda filum*), one of the brown algae which penetrates furthest into the Baltic Sea. In the Baltic, the importance of eel grass decreases with salinity and is completely absent to the north of the province of Uppland. However, several fresh water plants of the families *Potamogeton*, *Ruppia*, *Myriophyllum* and others take over and make up an increasingly large proportion of the vegetation the further north we proceed.

BOTTOM-LIVING MICROALGAE

Although most macroalgae are mainly lacking on the sediment bottoms, there is an abundance here of microscopically small algae with representatives of several hundred species among the same algal groups as in phytoplankton. The most common groups are diatoms and cyanobacteria (blue-green algae), but also common are flagellates, including dinoflagel-

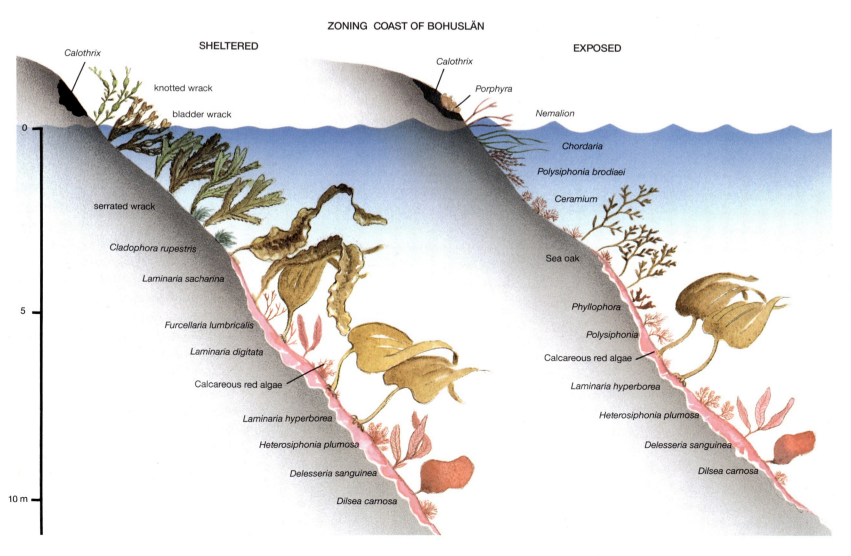

Perfoliate pondweed
(*Potamogeton perfoliatus*)

Stonewort
(*Tolypella nidifica*)

lates. Many of the diatoms live attached to grains of sand or can glide through the cavities between the sediment particles in numbers of more than a hundred individuals on one single grain of sand.

Sessile species of diatoms are also common as epiphytes on other plants, particularly on filamentous algae, which may become entirely covered causing them to assume a brownish colour. Diatoms can also grow directly on stones and rocks, particularly during the spring. However, owing to their small size they are generally not noticed until after a period of low water causing the dried-out dead shells of silica to shine like a white fringe. The bottom-living microalgae may also loosen and, aided by the oxygen formed during photosynthesis, float up to the surface in agglomerations that may have a size of more than a square metre.

CHANGES IN THE BOTTOM VEGETATION

Although the total production and turnover of nutrients in Swedish marine areas is dominated by phytoplankton, the bottom vegetation is of major importance. Large algae form the basis for the complex structure found in algal communities. Perennial species also contribute to the stability of communities by binding nutrients for long periods of time, whereas short-lived algae rapidly reach the end of their life cycle and loosen from the surface to become included in the mass of organic material decomposing on the bottoms.

The changed life style of Man has left its mark on the algal vegetation during the last century. Since the filamentous algae grow faster, they are favoured by the increasing amounts of nutrients reaching the sea. In an increasing number of areas, bladder wrack and its close relatives are being replaced by a border of green algae.

During the first decade of the 20th century, the algal vegetation along most of the Swedish west coast was surveyed by the renowned seaweed researcher Harald Kylin. Under-water studies conducted by Torsten Gislén on the hard bottoms of Gullmars fiord and other fiords have allowed several comparisons to be made during recent years, almost all of which have revealed larger sediment accumulations and a decreased vertical distribution of the algae. This has also been seen in comparisons between Mats Waern's studies in the 1950–60's and in 1989–90 in surveys made by researchers onboard the Uppsala-based R/V Sunbeam in expeditions along the Swedish west coast. In similarity with several other scientists, they found increasing amounts of filamentous algae in the southern Kattegat during recent decades, which, since the mid–1970's, have caused major problems for fishing and recreation. Continuous monitoring of algal vegetation in the Båstad area since the 1950's has been done by Tore Wennberg, Göteborg, who, continuously through 40 years, has followed the effects of eutrophication in the area by studying the composition of the algal vegetation.

There are several archipelago areas also in the Baltic Sea where bladder wrack has disappeared and has been replaced by filamentous algae. A follow-up of the renowned algal researcher Mats Waern's under-water investigations off Öregrund in the 1940's has shown that the vertical distribution of bladder wrack has moved up several metres during the last forty years on account of poorer light conditions.

Bottom Fauna

The animal species or groups living in and on the seabed are called the bottom fauna, or benthos. Sometimes the species living in the water immediately above the seabed are also called benthos (e.g., *Mysidae*). However, this is a rather incorrect description since they make daily migrations up through the water column to collect food.

BASIS OF DIVISION

The bottom fauna can be divided in several ways on the basis of their mode of life. Species living on the seabed are called epifauna whereas species living, at least mainly, buried in the bottom are called infauna. Epifauna can be divided into mobile and sessile groups whereas infauna can be divided into a number of sub-groups depending on the size of the animal: macro-, meio- or micro-fauna. The macro-fauna include animals which can easily be seen with the naked eye (in practice, the animals which are retained on a screen with a mesh size of 1 mm). Smaller animals are divided into meio-fauna (those which pass through the 1 mm screen but which get caught by the 0.1 mm mesh size) and the micro-fauna (animals passing through the 0.1 mm screen which cannot be seen by the naked eye). This division has been established for practical reasons and has no true biological relevance.

DISTRIBUTION AND ZONATION OF THE FAUNA

Species composition and number of species in the bottom fauna is influenced by factors such as the type of the substrate, salinity and many others. We distinguish between soft bottom fauna and hard bottom fauna. The soft bottom fauna, consisting of both epifauna and infauna, are found on bottoms which consist of fine gravel and sand (transport bottoms) or clay and mud (deposition bottoms). The hard bottom fauna, living on bottoms consisting of rock, boulders, stones or coarse gravel (erosion bottoms), consist mainly of epifauna.

The large variations in water salinity also have a major influence on the composition of the bottom fauna. In the northern and central parts of the Baltic, the fresh water and brackish water species dominate, whereas the fauna in the Kattegat and Skagerrak almost exclusively consist of marine species. Many species have their distribution limit at or close to the sill areas which restrict the exchange of water, e.g., at the Sound, the Åland Sea and the Quark.

The changed wave length of light and its decreasing strength with the depth of the water means that the flora gradually changes, which in turn influences the fauna. Consequently, the composition of the bottom fauna in shallow waters, which are reached by the light and where there is primary production (the phytal zone), is different from that on deep bottoms. On the deeper bottoms, the fauna must collect its energy from material which sediments down from the surface water.

In the same way as plants, the bottom fauna occurs often in distinct zonation patterns. The most important environmental conditions behind the emergence of zonation patterns are: the character of the bottom (colour, structure and hardness), the drying-out or degree of exposure (exposed to wave action) of the sea floor, and competition between species. The degree of desiccation is one important factor which results in the emergence of a zonation pattern whereas biological interactions such as competition influence the species composition within each respective zone.

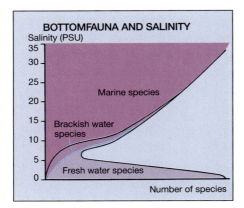

Salinity affects the composition of the bottom fauna. Large parts of the Baltic Sea are relatively poor in species. Few brackish water species have had time to develop during the relatively short history of the Baltic as a brackish sea.

The map is based on data collected during 1987–1990. (G54)

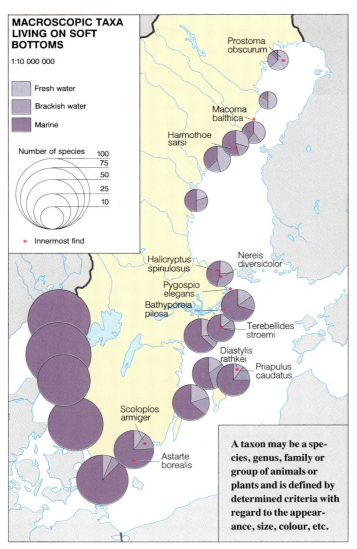

On the west coast of Sweden the balanid belt usually clearly marks the zonation of the flora and fauna.

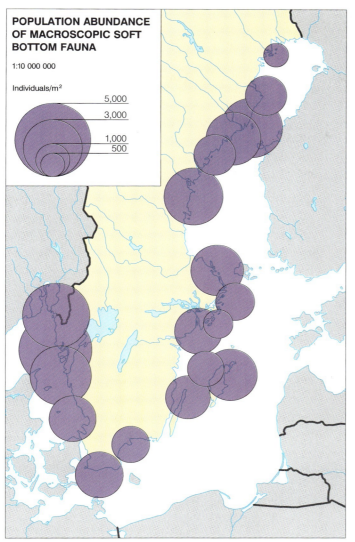

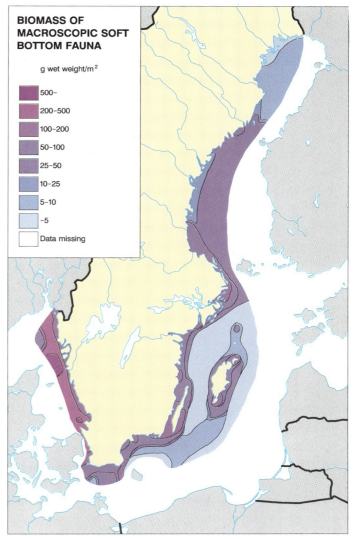

Nutrient-deficient conditions, together with low salinity and temperature, lead to the density of individuals being lowest in the Bothnian Bay.(G55)

Biomass is lowest in the oxygen-poor deep areas of the Baltic. Closer to the coasts, the biomass is higher than further out.(G56)

SHALLOW BOTTOMS OF THE BALTIC

On the shallow bottoms of the Baltic, we find microscopical algae and vascular plants. Hard bottoms covered with algae dominate. In sheltered places, the hard bottoms may turn into shallow soft bottoms.

Because of the dominance of hard bottoms, the fauna of shallow areas consists mainly of epifauna, which filter microorganisms from the water, graze on the epiphytes on stones and algae, or predate on other animals. Mussels (suspension feeders) and snails (grazers) of different kinds are dominating features here. In archipelagoes where small wave-exposed coves with soft bottoms can be found, we can also find infauna such as the mud-burrowing amphipod *Corophium volutator* and rag worms (*Hediste diversicolor*). Archipelagoes are more species-rich than open coasts.

The most important suspension feeders, i.e., common mussel, balanids and bryozoans are entirely absent in the Bothnian Bay. Instead, fresh water mussels are found in the innermost parts of the archipelagoes of the Bothnian Bay.

In the Baltic Proper, the common mussel frequently makes up 90% of the biomass on the shallow bottoms. Its dominance decreases strongly in the Bothian Sea, and it is completely absent in the Quark. The total biomass (amount by weight of living animals per unit area measured as g dry weight/m^2) is therefore greater in the Baltic Proper than further to the north. In addition, the total amount of biomass produced annually decreases the further to the north in the Baltic we get.

The amounts of crustaceans and snails decrease with depth whereas the dominance of mussels increases. On hard bottoms below the algal limit, the dominance of the common mussel is almost total.

The fauna on shallow bottoms is frequently dominated by mussels.

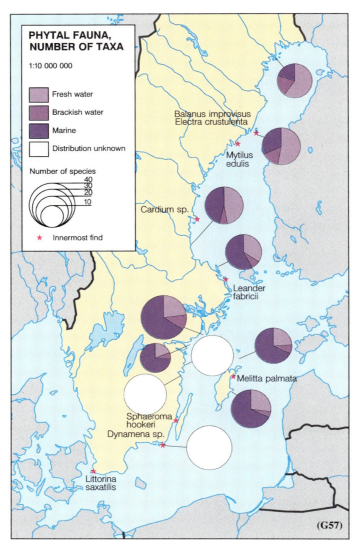
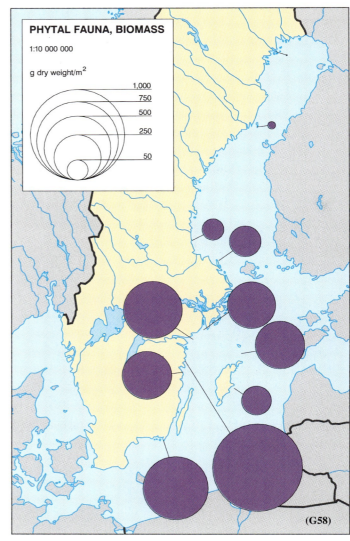

The number of species/animal groups in the phytal zone is lowest at exposed places such as Källskären, off Oxelösund, or in the low saline Bothnian Bay.

The low number or complete absence of common mussel is the main reason for the very low amount of animals in the phytal zone of the Gulf of Bothnia.

THE ROCKS OF THE WEST COAST

On the hard bottoms of the west coast the animal and plant life often is extremely rich. The animals are often sessile or semi-sessile.

A hard bottom on the west coast of Sweden.

On rocks in and above the water line, we often find the yellow lichen *Xanthoria*. In the splash zone the rocks are usually covered by the black lichen *Verrucaria*. Sometimes this lichen cover is replaced or mixed with blue-green algae. Periwinkles (*Littorina saxatilis*) and certain terrestrial animals (e.g. insects) are also common. Beneath this belt, there is a sharply delimited belt of balanids. Balanids are extremely specialized crustaceans which envelop themselves with a shell of limestone consisting of numerous individual sections.

The balanid belt also includes the common mussel (*Mytilus edulis*) which competes with balanids for space. Below the balanid belt, there is a belt of calciferous algae in which the first brown algae can be found.

Other important animals found in the shallow areas are, for example, the common starfish (*Asterias rubens*), sea urchins (e.g., *Echinus esculentus, Echinus acutus*), and the dog-whelk (*Nucella lapillus*). Starfish are often present in large amounts together with the common mussel, which is their favourite food. Sea urchins are considered to be among the most important herbivores (plant-eaters) in the hard bottom ecosystem. On some occasions sea urchins have been observed to totally destroy entire communities of brown algae. The dog-whelk is often found together with balanids. It has a strong radula with which it penetrates the shell of the balanids in order to eat the crustaceans. Mobile crustaceans are also found, such as shore crabs, spider crabs, edible crabs and lobsters.

THE SANDY SHORES OF THE WEST COAST

Most animals living along the sandy shores of the west coast belong to the infauna. Among the relatively few animals which spend most of their life on the sediment surface (epifauna), mention can be made of the shrimp *Crangon crangon* and the snail *Hydrobia*. Macroalgae rarely occur on this type of bottom, but instead the eel grass *Zostera marina* is frequently common. The meiofauna is very numerous, including, for example, groups such as *Nematoda, Kinorhyncha, Gastrotricha* and *Tardigrada*.

Small crustaceans are also common.

Among the infauna, the lug worm *Arenicola marina* is probably the one that is most noticeable in shallow soft-bottom areas. This worm, which lives in a deep U-shaped burrow in the sediment, produces characteristic heaps of excrement on the surface of the sediment. On sandy bottoms we find not only lug worms but also burrowing crustaceans and large numbers of molluscs. Some of the most common bivalves are the edible cockle (*Cerastoderma edule*) and the sand gaper (*Mya arenaria*). Certain burrowing sea urchins (e.g., *Echinocardium*) are also found.

The meadows of eel grass also provide cover for numerous animals within the epifauna group. Shore crabs (*Carcinus maenas*) and starfish (e.g., *Astropecten*) are a couple of examples.

As on hard bottoms, the animals within the shallow soft bottom systems occur in zones. Owing to the slight slope of the bottoms, however, these zones are more diffuse than those normally applying to hard bottoms.

DEEP BOTTOMS OF THE BALTIC

Soft bottoms dominate completely at greater depths in the Baltic. Of the total area of the Baltic with depths greater than 20 m, the proportion of hard bottoms is negligible. For the entire depth range between 4 and 50 m they make up less than 16% of the area.

Fauna living on soft bottoms are dominated by infauna. The method of feeding varies. Deposit feeders, which live on organic material in the sediment (including bacteria and other microorganisms which live there) make up the largest animal group, followed by suspension feeders and predators.

Microfauna as dominating animal group is called ciliates. They are unicellular animals which live in the pores and cavities in the uppermost layer of sediment on the seabed. In the Askö area of the northern Baltic Proper there may be as many as 17 million/m^2 but with a biomass of only 0.3 g dry weight. The microfauna largely constitute the food base for meiofauna and, despite its negligible biomass, make up 14% of the entire turnover rate of the bottom fauna.

The meiofauna in the same way as the microfauna, live in pores and cavities of the sediment. The dominating animal group is roundworms (Nematoda). Other important groups are copepods (Harpacticidae) and mussel shrimps (Ostracoda). They are usually found in numbers ranging between 0.5 and 4 million/m^2 with a biomass of 5–20 g wet weight. In comparison with the macrofauna, the meiofauna makes up a larger share of the biomass in the Bothian Bay, the ratio is about 2.5:1. In the Bothnian Sea and the Baltic Proper, the importance of meiofauna for biomass decreases successively. The ratio decreases to 1:10 in the Bothnian Sea, and 1:20 in the Baltic Proper. Even if the meiofauna biomass is only one-tenth of the macrofauna biomass, it is responsible for almost 30% of the entire turnover rate of the bottom fauna.

The macrofauna is the part of the bottom fauna about which we know most. In most of the Baltic, it is dominated by crustaceans such as the amphipods *Pontoporeia* and *Corophium* or mussels such as the Baltic tellin (*Macoma*). On some of the deeper bottoms in the Baltic Proper, the bristle worms make up the largest individual group of animals.

The fresh water species dominate in the Bothnian Bay, whereas species of marine origin dominate in the Baltic Proper. The Bothnian Sea is a zone of transition. The number of species/animal groups decreases from south to north.

The number of individuals is lowest in the nutrient-poor Bothnian Bay. In the Quark and the Bothnian Sea, there are often large numbers of individuals, mainly of the small amphipod Pontoporeia. In the Baltic proper, where the larger bivalves are an important feature, the number of individuals is slightly less.

The biomass is also lowest in the

PRIMARY AND MACROFAUNA PRODUCTION (KJ PER M^2 AND YEAR)		
	Primary production in the pelagic zone	Macrofauna production
Luleå skärgård (Bothnian Bay)	500	30
Holmöarna (Quark)	--	90
Norrbyområdet (Northern Bothnian Sea)	2 500	160
Southern Bothnia Sea	3 000	160
Asköområdet (Northern Baltic Proper)	6 000	220
Kiel Bay (Southwestern Baltic Proper)	9 200	440

The smallest macroscopic animals (0.1–1 mm) belong to the meiofauna. From the left, a roundworm, a copepod and an ostracod.

The apparently dead soft bottoms of the Baltic Proper are the homes of several organisms. The cod is approaching some mysids or opossum shrimps (*Mysis relictus*). In the bottom sediment can be seen, from left to right, a brush-worm (*Terebellides stroemi*), the Baltic tellin (*Macoma baltica*), a priapulid (*Halicryptus spinulosus*), a brush-worm (*Harmothoë sarsi*) and an amphipod (*Pontoporeia affinis*).

north and increases towards the south. This is both an effect of the water being richer in nutrients further to the south and that the share of bivalves, which have a large individual weight, increase towards the south. In addition, the biomass usually decreases with depth since the nutrient content of the sedimenting particles is poorer at greater depths. This results in the biomass being higher at the coast than further out to sea. The poor oxygen conditions in the deep areas of the Baltic Proper (below the halocline) lead to very low biomasses. In areas with hydrogen sulphide, the bottom fauna is completely absent and the bottoms are "dead".

DEEP BOTTOM COMMUNITIES OF THE SKAGERRAK/KATTEGAT

As also is the case in the Baltic, the deeper areas in the Skagerrak and Kattegat consist mainly of soft bottoms. The infauna dominate. The Danish biologist C.G.J. Petersen carefully described a number of bottom communities in the Skagerrak/Kattegat during 1911–1918 and grouped them according to their characteristic species composition. These are: Shore fauna, Macoma community, Syndosmya community, Venus community, Amphiura community, Haploops community, Maldane-Ophiura sarsi community and the Amphilepis-Pecten vitreus community. The shore fauna and Macoma community (named after the small Baltic bivalve) have been discussed earlier. The Macoma community is generally found extending from the shoreline down to about 8–10 m deep.

The Syndosmya community is named after the small white bivalve *Syndosmya (Abra) alba* which may occur in very large numbers. This community is also called a fiord community and is found down to depths of about 30 m.

The Venus community, or the North Sea community, is named after another small white mussel named *Venus gallina*. This mussel is found on relatively exposed bottoms with features of sand and may be found down to depths of about 40 m. Several other characteristic mussels are also found, e.g., *Spisula* and *Tellina*.

The Amphiura community is found on soft bottoms from about 20 m down to more than 100 m. This community is named after the brittle star *Amphiura*. Individually, the brittle stars dominate here and only a small number of bivalves are found. The irregularly burrowing sea urchin *Brissopsis* is also commonly on these bottoms together with burrowing crustaceans.

In the Haploops community neither bivalves nor brittle stars dominate but two small amphipods of the genus Haploops. These amphipods build small tubes which emerge 0.5–1 cm above the bottom sediment in very large numbers. The Haploops community is usually found on soft bottoms at depths 20 to 40 m.

Both the Maldane-Ophiura sarsi community and the Amphilepis-Pecten vitreus community are not found until we reach great depths. Both communities therefore only occur in the Skagerrak. The Maldane-Ophiura sarsi community, or the Skagerrak community as it is also called, is found at depths below about 150 m and got its name from the bristle worm *Maldane sarsi* and the brittle star *Ophiura sarsi*. The Amphilepis-Pecten vitreus community is not encountered until depths of 400–700 m. It is named after the brittle star *Amphilepis norvegica* and the scallop *Pecten vitreus*.

The Syndosmya community is one of the bottom communities of the Skagerrak/Kattegat and is found at depths of about 30 m. From the left, the species are the bivalves *Syndosmya (Abra) alba* and *Nuculoma tenuis*, a common whelk (*Buccinum undatum*), a trumpet worm (*Pectinaria koreni*), a bivalve (*Corbula gibba*), a blunt gaper (*Mya truncata*), a brittle star (*Ophiura texturata*), and a razor shell (*Cultellus pellucidus*). The shaded parts of the illustration show oxygen-free sediment.

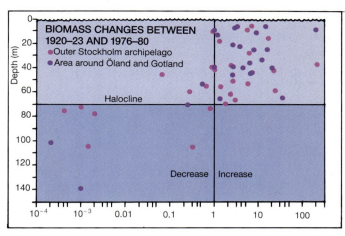

During 1976–1980 considerably higher benthic biomass was found above the halocline in the Baltic Proper, compared to the period 1920–1923.

An overview of the amount of benthic fauna and fish, given in thousands of tonnes, in the Kattegat in 1957.

ROLE OF BOTTOM FAUNA IN THE ECOSYSTEM

A large part of the organic material produced in the surface water by algae and bacteria will sediment to the seabed or is taken up by suspension feeding bottom organisms. Most of this organic material is broken down (remineralized) whereby the nutrients carbon, nitrogen and phosphorus are released which, by means of the movements in the water, can be transported up to the surface water where they once again can be used for new organic production.

The bottom fauna participates directly in this recirculation by means of the degradation that takes place when food passes through the intestine. In addition, the fauna contribute indirectly to the degradation through their physical activity in the soft bottoms. The burrowing forms pump oxygen-rich water down to deeper levels in the bottom sediment where the oxygen would otherwise not reach.

The bottom fauna also makes up an important source of food for bottom-living fish species. In the Kattegat, 6–7% of the "edible" bottom fauna's biomass is metabolized in fish biomass. The corresponding figure for the entire bottom fauna is about 1%. Since the Baltic mainly lacks deep-burrowing and large-growing species of bottom fauna, a considerably larger part of the bottom fauna can be used as fish food. The size of this share is not known but is probably much larger than in the Kattegat.

CHANGES IN TIME

Seasonal variations. During the period of the year when the bottom fauna produce a new generation of larvae and young, the individual density (abundance) and weight (biomass) of the bottom fauna increases. Depending on the water depth, this takes place from early spring to late autumn. During the rest of the year, when generally no recruitment takes place and food availability is poorer, then predation from fish and other predators, together with natural mortality, usually leads to the volume of bottom fauna decreasing to a minimum during the late autumn/winter.

Inter-annual variations. Depending on variations in ambient factors (e.g. salinity and temperature of the water, or changed primary production and sedimentation) between different years, the abundance and biomass of the bottom fauna will change. In specific localities, the inter-annual variations may be large. However, if larger areas are studied, then variations are much smaller. In the case of the soft bottoms in both the Baltic and at the west coast, the inter-annual variations in biomass are normally less than a factor two.

Long-term changes in temperature, salinity, oxygen content and contents of nutrients and environmental toxins in the sea may cause long-term changes in the bottom fauna. Increased temperature will increase the growth of the animals, decreased salinity and oxygen contents will lead to the decimation of certain species. An increase in nutrients will lead, via increased plankton production and sedimentation, to increased production and thereby an increased biomass of bottom fauna. If the amount of nutrients increases still further, then the species composition will be changed and ultimately the fauna will die as a result of the oxygen deficiency which will occur when excessive quantities of organic material are to be degraded.

The increase in salinity which took place in the Baltic during the 1970's led to both immigration of species from the Belt Sea as well as the distribution boundary being moved further to the north for certain species.

During recent years, reports of increased amounts of bottom fauna and sometimes even modified species compositions have been received from many areas. Particularly in coastal areas, with their coves and archipelagos, the changes have been considerable. However, changes have also taken place in the open sea. In both the Baltic and in the Kattegat and Skagerrak the biomass has increased in comparison with the situation earlier. In the Baltic, however, the bottom fauna has largely been obliterated from areas below depths of 100 m as a result of the increased oxygen consumption below the halocline and the poorer exchange of bottom water during the 1980's.

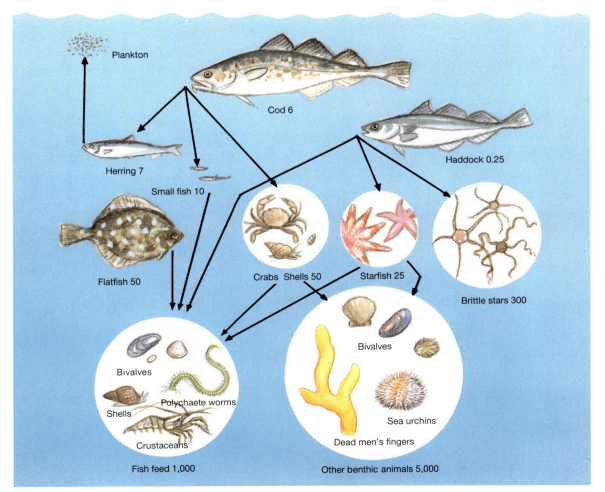

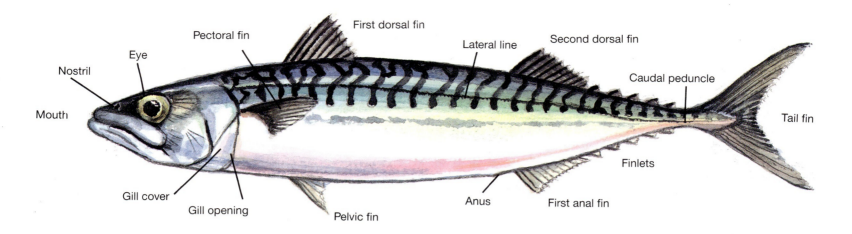

A typical fish has fins and a streamlined shape. Cod, herring or, as shown here, mackerel, are good examples. Their shape allows them to move rapidly through the water with a minimum of resistance.

Fish

A typical fish has fins: a caudal fin and dorsal and anal fins, pair-wise pectoral and pelvic fins. Fins are folds of skin which are supported by rigid or soft rays. They are flexible and can be turned and folded. Fins are essential to the swimming ability of fish: They function as stabilizers, as bow propellers, brake flaps and as rudders for vertical and lateral control.

On most bony fishes, the body is covered by thin scales whereas sharks, etc., have denticles. The scales grow at the same rate as the fish and are therefore a valuable aid when determining age.

A typical characteristic of fish is that they respire through gills. The gills consist of lamellae that are attached to cartilaginous gill arches. They are thin-walled folds of skin with a rich blood supply and are capable of taking up some of the oxygen dissolved in the water and releasing carbon dioxide and other combustion products. In plankton-eating species, the gills are also equipped with gill rakers, a sieving mechanism whereby fish can separate food organisms from respiratory water.

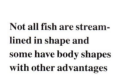

Not all fish are streamlined in shape and some have body shapes with other advantages than fast swimming.

Fish respire by means of their gills, whereby they absorb oxygen dissolved in the water.

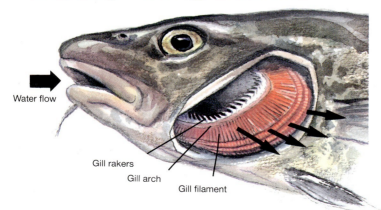

SWIMBLADDER

The swimbladder develops as an extension of the gut. Some species retain this contact with the foregut in the form of a narrow connection. Such species, when fry, fill their swimbladder by going up to the surface and swallowing air. In most fishes the swimbladder, on the other hand, is closed and regulates the amount of gas through the gas glands, which transfer gas between the swimbladder and the blood. By altering the amount of gas in the bladder, the fish adapts its specific weight to the water pressure. The deeper the fish goes in the water, the more gas it must transfer to the swimbladder in order to retain buoyancy. Without a swimbladder the fish has a density of about 1.06–1.02 g/cm^3 and thus sinks when it stops swimming. Several fish lack swimbladders, e.g., mackerel, flatfish, and sculpins. It is believed that mackerel and tuna more easily escape from toothed whales, such as dolphins and killer whales, as a result of not possessing a swimbladder. These whales use sound impulses (echolode) to localize their prey and it is largely the swimbladder which causes the echo made by a fish.

In some fish the swimbladder has developed a hearing function: herrings have a couple of extensions of the swimbladder up to the inner ear whereas carp fishes have developed a series of small bones which transfer sound waves from the swimbladder to the inner ear. The swimbladder has also been utilized to create sound. The gurnard makes its sounds by vibrating the swimbladder.

The function and appearance of the swimbladder varies between species, with some fish completely lacking a swimbladder.

SENSE ORGANS

Organs of taste may be located almost anywhere on the body in fish: In the mouth, on the lips, on the body and, not least, on the barbels.

A sense organ which is characteristic of water-living animals is the lateral line. Small sensory organs (neuromasts) which are capable of registering changes in the movements of water and pressure are located in a canal along the sides of the body of the fish and in a system of branched canals on the head. Thus, the movement from a polychaet creeping on a muddy bottom can be registered by the lateral line even if the worm is not visible. Other situations when the lateral line provides important information on the surroundings of the fish is in providing early warning of approaching predators and the localization of invisible obstacles in the water.

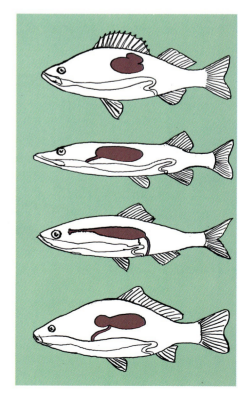

TOO MUCH/TOO LITTLE SALT

Fish have a concentration of dissolved salts in their body fluid of about 0.9%, i.e., it is higher than in freshwater but lower than the salinity of seawater. Nature attempts to level out the salinity concentrations in the two fluids which come into contact with each other. Fish in both saltwater and freshwater must actively counteract this process. In saltwater the body fluid tends to leach out or be drawn out of the body. Fish drink saltwater in order to compensate for the loss of fluid and have special cells on their gills and in the intestine which they use to remove the surplus salt.

In freshwater the problem is the reverse. Water enters the body through the skin and tissues. The fish does not need to drink but has to remove surplus water. Consequently, they produce large quantities of urine which has been desalted by the kidneys. The remaining salt requirement is provided by the gills which absorb salt from the respiration water.

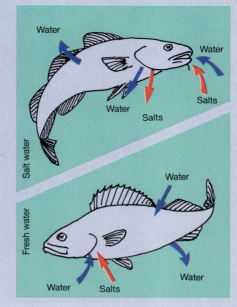

REPRODUCTION STRATEGIES

There are many different ways of obtaining as large a proportion of surviving progeny as possible, and the same also applies to fish. Most marine fish produce a large number of small eggs which are released and fertilized in the open water (pelagic eggs). Both flatfish and cod use this strategy: a plaice releases about 0.5 million eggs, a cod 1–5 million, and a ling 20–50 million. The number released by a sun-fish is almost astronomical – 300 million eggs! It is clear that such a cloud of high-quality protein is regarded as a welcome addition to the nutrition of many other animal species and, indeed, they make the best of the opportunity. The large number of eggs, and the dispersal and thinning-out of the egg mass which occurs very rapidly, contribute to a sufficient number surviving. It may be quite enough if only 1% of the eggs survive.

For fish which lay their eggs on the bottom (demersal eggs) the probability for survival of each egg is greater; these therefore need fewer eggs. Herring, which lays its eggs on gravel or on plants, releases 20,000–50,000 eggs, whereas sandeels, which place them on/in the sandy bottom, lay up to 20,000.

Instead of releasing large quantities of protein in the form of eggs, each of which has little of survival, there may be better chances of survival for a smaller number of eggs. Thus, for example, the three-spined stickleback glues together plant parts to form a nest, in which the male often persuades several females to lay eggs. He then protects his collective clutch (300–1,000 eggs): he defends it, flushes it with fresh water with his pectoral fins and removes dead eggs.

The sand goby (found in shallow shores) also protects its eggs but uses empty mussel shells, on which the eggs are attached. A very specialized form of egg care has been developed by the pipefish where the male carries the eggs in a special pouch on its belly.

Fish belong to the oldest vertebrate animals and today number more than 20,000 species, of which 13,000 live in the sea. In Swedish waters there are 190 fish species, of which about 35 are rare and found only at the West Coast.(G59)

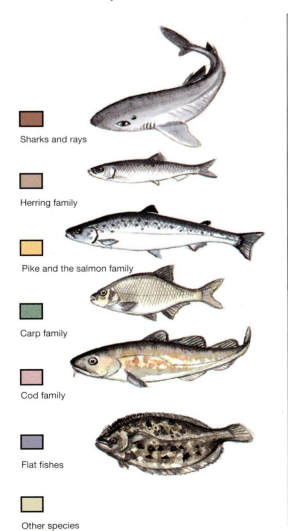

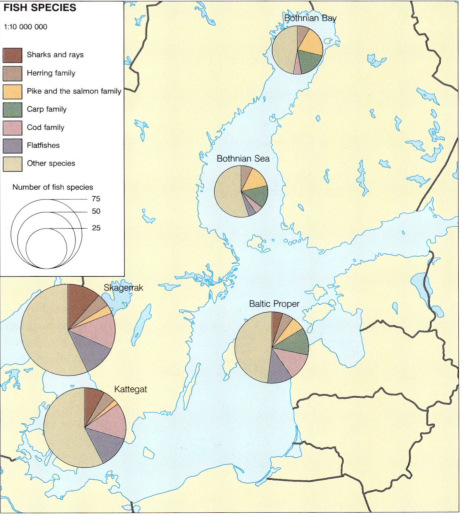

Marine Resources

The importance of fishing in coastal communities can still be seen today in many places, such as here in Smögen harbour.

Sweden has a long coastline, about 7,000 km, and has access to a marine area which is about 40% of the land area. In the Baltic, Sweden is the largest shoreline state and therefore has special responsibility in international cooperation on the marine environment. Most of Sweden's population live in coastal areas. For many of these areas, the marine industries – fishing, shipping and ship-building – have been a feature of the history and livelihood of the coastal areas for hundred of years. For natural reasons, contact with the sea has been, and still is, concentrated to marine areas close to the coast. It is also in such areas that pressure on the utilization of marine resources and the threat to the marine environment is most noticeable.

Marine resources are usually grouped according to different ways of utilizing the sea. Resources which mainly come to mind are perhaps fish, the sea as a means of transportation, and seabed resources. The International Convention on the Law of the Sea gives a much more detailed account of ways of utilizing marine resources and lists them under the following main sectors:

- Food production
- Extraction of minerals
- Energy exploitation
- Transports
- Military use
- Use of water areas for effluents
- Outdoor recreation
- Marine historical preservation
- Marine nature conservancy
- Building activities
- Research and technical development

Interest in the sea and its resources has grown considerably following the introduction of the International Convention on the Law of the Sea in 1982. Future development in the marine sector may be expected to lead to increased utilization of marine resources depending on how technology develops, etc. In turn, this is linked to economic and social development in different parts of the world.

The new interest in marine resources, which is partly connected with the discovery of large oil and gas reserves in the sea elsewhere, has so far not affected conditions in Sweden other than marginally. Interest in prospecting for oil and gas is strong today at several places in the Baltic. There are suitable geological conditions for finding oil and gas fields in Swedish marine areas.

Aquaculture should also be mentioned among the new interests. Production of fish in cages and mussel farms have become a permanent activity in different places, mainly in the archipelagos. Mussel farms are found only on the west coast. Swedish aquaculture is mainly conducted on a small scale and is growing slowly.

In addition to the above, the function of marine areas around Sweden for recreational purposes has become increasingly important. Such areas are of greater importance for the Swedish people than most of them realize. It is estimated, for example, that more than two million people visit coastal establishments every summer.

The marine question which has attracted most attention in Sweden during recent years is the problem of pollution related to the emission and discharge of various toxic substances and nutrients. Our large areas of sea are, nevertheless, not infinitely large and the mean depth no greater than taller buildings and towers would emerge above the water surface if we were to place our towns and cities on the seabed.

A future resource utilization which has now become of interest and which may be developed is sea-based wind energy. Several places off the coasts of Sweden have been considered suitable for the siting of groups of larger wind energy plants, 3 MW and larger.

The threat to the marine environment also requires protection for the different ecosystems in the sea. Pressure on the utilization of such areas needs equally to decrease. Nature conservation and historical preservation of marine areas will therefore become increasingly important.

Knowledge of the sea – both social, scientific and technical – is still, how-

The economic shares of different industries in marine industrial activities.

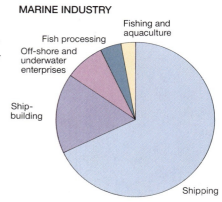

ever, fairly undeveloped compared with corresponding knowledge of terrestrial aspects.

The environmental problems of the sea, both our own and others, as well as changes in the sea, make demands for continued improvement of our basic knowledge by means of marine research and environmental monitoring.

The increased use of marine resources will also lead to requirements for a developed system in order to be able to handle different demands and to solve conflicts. A system of this kind must be based on knowledge of, e.g., administrative boundaries, laws and regulations within different sectors, different interests and environmental conditions, as well as on effective surveillance of the utilization of marine resources and the environment. Overall physical planning should be developed into the coordinated instrument necessary in a system of this kind for a more planned approach to the use and protection of marine resources.

INTERNATIONAL CONVENTION ON THE LAW OF THE SEA

Which, among other things deals with the rights of coastal states to:
- territorial waters of 12 nautical miles
- economic zone of 200 nautical miles
- to manage seabed resources.

Aquaculture has become increasingly common in many places around Swedish coasts. Fish-farming using cages near Kungshamn, Bohuslän.

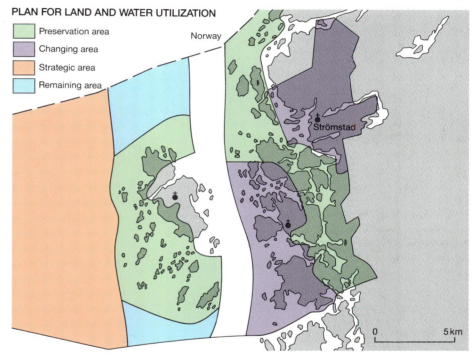

It is important to know what areas of land and water are being used for, as well as the environmental situation in coastal regions. Here is an example of a synoptic plan for a coastal water area in Strömstad municipality.

Clean water along beaches and in archipelagoes is of great importance for recreation and leisure activities.
Waiting for the coffee to boil on the coast of the Gulf of Bothnia.

Boundaries of the Sea

The Swedish National Board of Fisheries

Coast Guard

University of Göteborg

Legally, the sea is divided into inner waters, territorial waters and "the open sea". There are also additional zones, fishing zones and economic zones. Inner waters include not only lakes, canals and rivers, but also areas of water within and between headlands, islands, islets and skerries. The boundary between inner waters and the territorial waters consists of the base-line. The open sea starts at the territorial limit, where the territorial waters end.

Additional zones, fishing zones and economic zones are made up of marine areas outside the territorial limit. Also within additional zones, the coastal state has the right to take action against infringements within its territory of health, immigration, environment, taxation and customs regulations. Within the fishing zone the coastal state has its "ownership right" to the living sources of the sea. In the economic zone, the coastal state has the sole right to utilize both living and non-living resources both in the sea and beneath the seabed.

In accordance with the 1982 International Conference on the Law of the Sea, Sweden has introduced a territorial limit of 12 nautical miles. In cases when this has not been possible a boundary agreement has been reached which is generally based on the mid-line principle. This states that the boundary between two states neighbouring each other will consist of a mid-line on which each point is situated at similar distances from the base-lines of each respective country.

The Swedish fishing-zone has been established by means of the mid-line principle in boundary agreements with our neighbours. Only a smaller area between Sweden and Finland to the southeast of the Åland Sea has still not been agreed upon (1992). In principle, this zone comprises the Swedish part of the continental shelf (as defined in the 1982 International Conference on the Law of the Sea). Sweden has still not introduced an economic zone. If this is done it will, in principle, follow the same boundary as the fishing-zone.

Swedish marine territory, i.e., inner and territorial waters, is divided from the fisheries viewpoint into private and public waters. Private waters include inner areas where the owner of the shore has the sole right to fishing, whereas in public waters the State "owns" the fishing rights. Since the coastal communes have successively been given increased responsibility for physical planning of neighbouring territorial waters, the division of territorial waters among communes has also started.

Institutes and Authorities

A large number of authorities, university departments and companies throughout Sweden are working with activities involving the sea. The list given below only includes major activities being conducted at national authorities and institutes with national or regional responsibilities.

STRÖMSTAD:

Tjärnö Marine Biology Station, a field station for the Universities of Göteborg and Stockholm.

LYSEKIL:

The Institute of Marine Research of the National Board of Fisheries makes population estimates of commercially important fish species.

FISKEBÄCKSKIL:

Kristineberg Marine Biological Station (Royal Swedish Academy of Science) conducts research and education. Part of the Göteborg Center for Marine Research is also located at the station. The Klubban Station, one of the field stations of the University of Uppsala, is nearby.

GÖTEBORG:

The head office of the National Board of Fisheries is responsible for the future development of fisheries in Sweden. The University of Göteborg has numerous departments working with marine research and education. Sweden's only department for physical oceanography is found here. Some of the activities are organized within the Göteborg Center for Marine Research. The Oceanographical Laboratory of the Swedish Meteorological and Hydrological Institute works with research, investigations and monitoring of the marine environment.

LUND:

Several departments at the University of Lund work with research and education in marine sciences.

KARLSKRONA:

The Coast Guard are responsible for patrolling the boundary and fishing-zones and dealing with oil and chemicals discharged at sea. The Institute of Marine Research at Lysekil also has a branch station here.

KALMAR:

Kalmar University has a marine ecology unit which is part of the Stockholm Center for Marine Research. Education, environmental investigations and monitoring are the activities conducted here.

GOTLAND:

There are two field stations, at Slite and Ar, which belong to the Department of Systems Ecology at the University of Stockholm.

LINKÖPING:

At the University of Linköping there are several departments working with research and education in marine sciences. Some of these are grouped under the common title Theme Water.

NORRKÖPING:

The National Administration of Shipping and Navigation conducts, e.g.,

hydrographical surveys at sea, and produces the Swedish charts. The Swedish Meteorological and Hydrological Institute's head office is also at Norrköping, where e.g. weather, ice, current and wave forecasts for shipping are produced.

STUDSVIK:

The Swedish Environmental Protection Agency has a laboratory here for brackish water toxicology.

ASKÖ:

The Askö Laboratory is a field station belonging to the Stockholm Center for Marine Research. Research and environmental monitoring are conducted here.

STOCKHOLM:

The University of Stockholm has a number of departments working with marine research and education. Some of the activities are organized within the Stockholm Center for Marine Research. The Swedish Environmental Protection Agency is responsible for, e.g., coordination of the environmental monitoring of Swedish marine areas. At some of the university laboratories, analyses are made of environmental toxins in, e.g., fish samples. The Swedish Museum of Natural History is, e.g., a centre for seal research, and also houses an environmental sample bank.

UPPSALA:

At the University of Uppsala there are departments which partly work with marine questions. The Swedish Geological Survey produces maps of the seabed's content of sand and other mineral resources.

ÖREGRUND AND FORSMARK:

The Coast Laboratory of the National Board of Fisheries is at Öregrund and is responsible for investigating coastal populations of fish.

UMEÅ:

At the University of Umeå, there are a number of departments which partly work with marine questions in their research and education. Some of the activities are organized within the Umeå Center for Marine Research.

NORRBYN:

At Norrbyn there is a field station which belongs to the Umeå Center for Marine Research.

Fisheries and Aquaculture

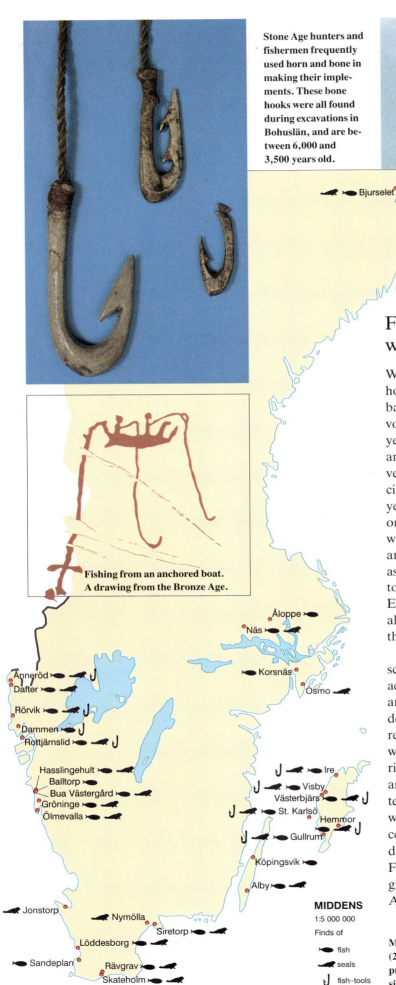

Stone Age hunters and fishermen frequently used horn and bone in making their implements. These bone hooks were all found during excavations in Bohuslän, and are between 6,000 and 3,500 years old.

SWEDISH CATCH OF HERRING SINCE THE MIDDLE AGES

Fishing from an anchored boat. A drawing from the Bronze Age.

MIDDENS
1:5 000 000
Finds of
- fish
- seals
- fish-tools

Most finds of fish are from the Late Stone Age (2,500–2,000 years B.C.). The map in fact provides more details about the actual excavation situation than about the distribution. (G61)

Fisheries – an Industry with Ancient Traditions

With spears, bows and arrows, with hooks and nets, with woven cages, barriers and traps, mankind has devoted his energies for thousands of years to catching fish and other sea animals. Ancient middens have revealed that more or less the same species were caught then, up to 10,000 years ago, as those caught today. It is only from one period with warmer sea water at about 2,000 B. C. that there are finds of more exotic species such as the bogue (*Boops boops*) which today is found to the south of the English Channel; during this period also the Greenland seals extended their migrations to Swedish waters.

Initially, fishing was on a small scale and unspecialized. It required access to salt, good transport facilities and, foremost, a large demand in order to benefit from the mass occurrences of fish. All these conditions were fulfilled during the medieval period when, starting in the 11th century and lasting for about 500 years, an intensive herring fishery was developed, with Skanör and Falsterbo as main centres. The fisheries were largest during the 14th and 15th centuries. Fishing was done with drift nets and gill nets during the period from August to October. Available reports on catches vary between 10,000–30,000 tonnes per season. However, these were not professional fishermen but burgers and farmers from Skåne and seafarers of various nationalities who arrived and took part in the fishing during the season. It is reported that 37,500 people were involved in the fishing during the 1520's. During the same period, the number of boats was reported to be 7,515. Herring was of major importance in trade policy and fishing was regulated with numerous rules and customs tariffs, which were supervised by Danish and Hanseatic officials.

Also in Bohuslän, the herrings periodically gave rich fishing. This took place when the over-wintering North Sea herring came so close to the coast that it could be caught with the nets and seines of that time. Three herring periods are well documented: From 1556 until 1587, during the years 1747–1809, and in the late 19th century (1877–1906). During the 16th century, gill nets were mainly used but during the 18th century were largely replaced by beach seines.

These large quantities of herring were used either for salting or for pro-

Herring fishery. Olaus Magnus, 1555.

duction of oil. In 1787, there were 338 salting houses and 429 train-oil factories. The train-oil was used for lighting; the street-lamps of Paris periodically used herring train-oil. Faster communications in the late 19th century resulted also in fresh herring being exported; in 1895/96 a total of 87,000 tonnes of herring was sent to Germany and England. At the same time, there were 23 guano factories where 14,000 tonnes of herring meal were made into fertilizers and 10,700 tonnes of oil.

During the years between herring periods the availability of coastal herrings was small, which led to difficult social and economic problems. Variations in hydrographical conditions and population fluctuations in herring populations of the North Sea are considered to be contributory reasons to these large changes in occurrence and availability of herring. It is important to emphasize that the decisive factor was that availability increased during the herring periods as a result of herrings coming closer to the coast. Rich abundances out at sea could not be utilized with vessels and equipment available at that time. This is something which can be done today. As soon as the herring population increases after one or more large yearclasses, the yield can be increased and modern equipment is used to search for the herrings wherever they may be. As a result, new herring periods no longer occur.

Swedish Fisheries During the 20th Century

Fishing efficiency has increased during the 20th century and today each fisherman catches, on average, 15 times more than could be caught at the turn of the century. The number of professional fishermen in Sweden has decreased by 81% since 1945, mainly in areas where fishing has been a dominating industry, i.e., the counties of Bohuslän, Malmöhus, Blekinge, Kalmar and Gävleborg.

Apart from a peak in the middle of the 1960's, the annual catch has varied around 200–250 thousand tonnes since the 1950's. The value in the first retailing stage of the total catch started to increase dramatically in the mid–1970's. However, this was only an effect of high inflation since if the value is expressed in fixed prices based on 1987, then there is a different picture. The average value per kilo caught, i.e., the price, was 1 SEK/kg in 1973 and in 1989 about 3 SEK/kg in current prices. A back-calculation to 1987 prices shows peaks during the war years in 1915–18 and 1941–45 but subsequently a slow decrease. The level today is among the lowest during the present century. If the total catch value is divided by the number of fishermen, then we get a picture of massively increasing income: From 20,000 SEK in 1964 to more than 200,000 SEK per person in 1989. Even when the calculation is based on a fixed monetary value (the cost of living index for 1987), the increase is large – an average income has increased two-fold during the period from 1964 to 1989.

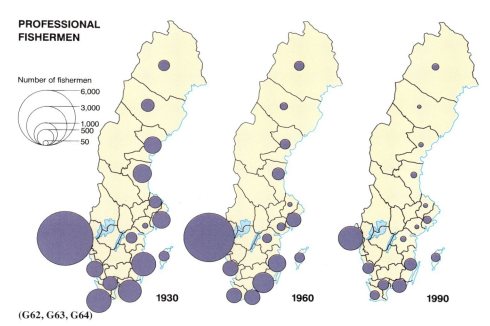

(G62, G63, G64)

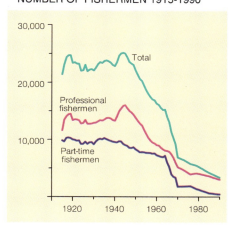

The number of fishermen has been calculated as the number of occupational fishermen plus half the number of people with fishing as a secondary source of income.

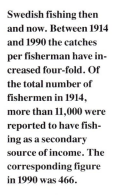

Swedish fishing then and now. Between 1914 and 1990 the catches per fisherman have increased four-fold. Of the total number of fishermen in 1914, more than 11,000 were reported to have fishing as a secondary source of income. The corresponding figure in 1990 was 466.

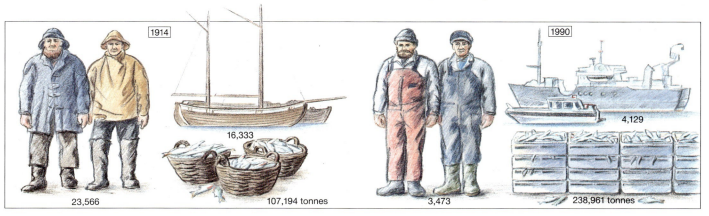

THE DEVELOPMENT OF SOME SWEDISH FISHERIES, 1929, 1949, 1969 AND 1989

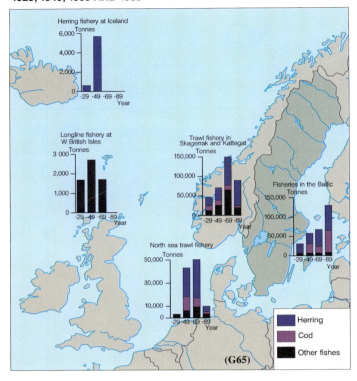

ICES – INTERNATIONAL COUNCIL FOR THE EXPLORATION OF THE SEA

ICES was founded in 1902. It has 17 member-nations: Belgium, Canada, Denmark, Finland, France, Germany, Iceland, Ireland, the Netherlands, Norway, Poland, Portugal, Spain, Sweden, the United Kingdom, the USA and Russia. Its secretariat is in Copenhagen.

ICES has the task of encouraging member-nations to conduct investigations and research on the sea as well as coordinating common research inputs, particularly with regard to living resources. The activities of ICES are concentrated in the north Atlantic, particularly the northeastern Atlantic and contiguous seas.

The large increase in catch per fisherman could be achieved as a result of technical development: Steam engines, crude oil engines and diesel engines have replaced oars and sails. Navigation aids have developed from the simple compass, through direction finders and navigators, to satellite-transmitted systems giving the vessel's position with great accuracy. The means of searching for fish have developed from hand-held plumb-lines, echo-sounders and sonar, to the acoustic systems of today which show on a colour monitor not only where the fish are located in the water column but also their length distribution, as well as providing an estimate of the quantity in tonnes.

The methods of handling the fish have progressed from placing them in woven baskets, through iced boxes, to freezing and storage on board in chilled sea-water.

This mechanization of fisheries places great demands on the knowledge of the fishermen, they must know not only where, when and how the fishing can be done best, but must also have a thorough technical training to be able to handle the equipment; they must have economic training and knowledge of legal matters in order to be able to run a company with a turnover amounting to millions of crowns.

POPULATION REGULATION

Living organisms of the sea belong to

The rise and fall of the fisheries. Excessive exploitation of fish resources results in decreased catches and a fishing fleet with unutilised capacity.

the category of finite but renewable resources. The fish we catch are replaced by new younger fish which gradually get larger. If fishing intensity is in balance with the amount supplied by the fish population in the form of individual weight increment and new young fish, then fishing can continue without the population being influenced. If catches are greater than supply, then the population will continuously decrease – it is being over-fished.

WHO OWNS THE FISH IN THE SEA?

Living resources in the sea are one of the few things which still can be regarded as a common resource: No single individual owns them as long as they remain in the water. This situation encourages over-exploitation.

Consequently, rules are required to establish a reasonable economy with the living resources. The migratory habits of fish imply that national rules are generally insufficient and international agreements must be entered into.

The present system for international fisheries management in the northeast Atlantic is based on member-countries annually asking ICES for advice on population utilization. The scientifically-based advice given by ICES on catch restrictions, mesh sizes, closed periods, closed areas and other regulatory measures, are then used as a basis for annual negotiations between individual nations, with the EC, and within fisheries commissions, concerning management of the common fish populations.

Regulation of fisheries usually occurs in such a way that the states concerned reach agreement on a total allowable catch (TAC) which is divided into national quotas. Attempts are also made to prevent young fish from being caught by means of regulations being imposed on the smallest permitted mesh size in trawls, seine nets and gill nets. In some cases there is a need to protect a population during spawning, for example, and then agreement is reached to introduce a ban on fishing within a certain area and/or during a certain season.

Sometimes, but by no means always, politicians follow the recommendations proposed by ICES. Frequently, they choose to place greater emphasis on socio-economic problems, and take a large catch even though this may lead to poorer yields in future years.

Catch areas for cod. In 1989 the world catch of cod was about 1.8 million tonnes. (G66)

Cod

Cod (*Gadus morhua*) is the most important bottom fish in the north Atlantic. It is found in both shallow waters and down to depths of about 600 m, usually close to the bottom, but it also hunts in the open water (pelagic).

Cod spawns in the spring in waters with a temperature of 4–6°C which, at the Swedish west coast, implies the period around March and in the southern Baltic in April–June; at places further north in the Baltic and in the Bothnian Sea these temperatures are reached during the summer, generally in July–August.

Cod produce a large number of eggs and the amount varies with the size of the female; from a few hundred thousand to a couple of million eggs. The eggs are pelagic, i.e., they float freely in the water and usually rise to the surface. In the Baltic, where the water is stratified with layers of different salinity, they will remain in a layer suitable for their buoyancy.

Cod may spawn in but do not reproduce in areas with low salinity, e.g., the Åland Sea or the Bothnian Sea. Replacement in such areas takes place by immigration of young cod from the south.

On the whole, the cod is omnivorous and feeds on all kinds of crustaceans, worms, molluscs and fish. The larger it gets the more fish it eats.

During the 1980's, cod has become increasingly important for Swedish fisheries and in 1989 made up almost 45% of the total catch value. This was a result both of occasional increases in availability of cod and on a changed emphasis in fisheries as a result of a modification in fishing policy. During 1978–83, recruitment to the catchable population went through a two-fold or three-fold increase, the population of adult cod increasing the most strongly. Cod fishing became attractive and total fishing inputs increased. Many vessels stopped fishing for herring and instead concentrated on cod.

Negotiations within the International Baltic Sea Fishery Commission (IBSFC) failed to result in any agreement on catch limits during 1982–88, partly depending on the large area of water between Gotland and the Baltic States which was in dispute and not divided between Sweden and the former Soviet Union until 1988. This area was regarded as international water and thus provided an opportunity for uncontrolled fishing. When recruitment to the population again fell to a lower level (year-classes 1984 and younger) and the number of adult cod again decreased, as well as the catches, this led to major problems for the excessively large fleets of cod-fishing vessels.

Regrettably, the agreements reached by IBSFC for total catches in 1989 and subsequently were so large that they have had no restrictive effect on international fisheries, and the reduction of the population is continuing. Massive in-flows of saline, oxygen-rich water from the Skagerrak/Kattegat will probably be necessary to provide a sound basis for continued large-scale cod-fishing in the Baltic.

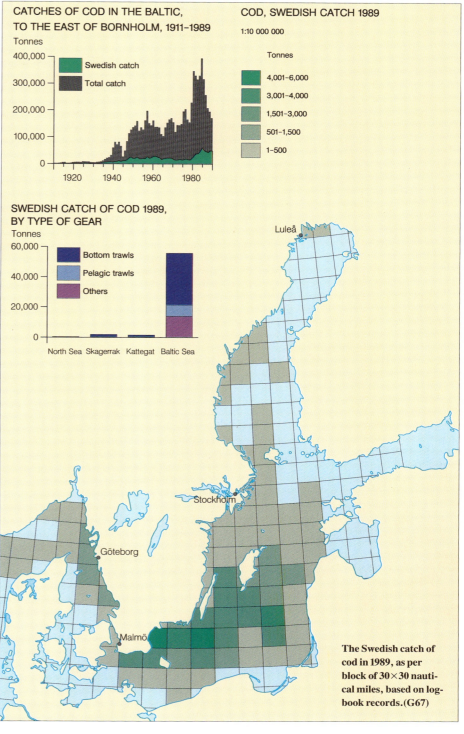

The Swedish catch of cod in 1989, as per block of 30×30 nautical miles, based on logbook records. (G67)

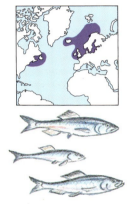

Catch areas for herring. In 1989 the world catch of herring was about 1.6 million tonnes.(G68)

"The whole world is acquainted with the herring. Everybody has at least on some occasion become acquainted with this useful fish which, when alive, sets thousands of arms in motion and, when dead, even more mouths". (Ch. Hagdahl, Kokkonsten).

Herring

The herring (*Clupea harengus*) is a pelagic species which may occur in enormous shoals. This species belongs to the *Clupeidae* family, together with sprats, sardines, anchovies and anchoveta, all of which are important for fisheries in different parts of the world.

Herrings are divided into a number of populations (also called stocks, races) which are distinguished by, e.g., different spawning times, migration routes, growth rates and sizes. The herrings which have been of greatest importance for fisheries are the Atlanto-Scandian herring and the North Sea herring.

The Atlanto-Scandian herring have been subjected to intensive fishing for many centuries. It was not until the 1950's, however, that new technology and a larger market (fishmeal and oil) provided opportunities for massive catches. At that time the catch was about 1,000,000 tonnes of Atlanto-Scandian herring annually up to 1966, when the catch rose to 2,000,000 tonnes. What had previously been the richest herring population in the world then collapsed, and has still not recovered.

A similar fate was met by the North Sea herring: after a few peak catches in the early 1960's with more than 1,000,000 tonnes annually, the population size and the catches have decreased steadily. A ban on fishing was introduced during 1978–1983. Subsequently, the North Sea herring has recovered and the population now permits an annual catch of about 500,000 tonnes.

On the west coast of Sweden the herring catch consists of a mixture of different populations. Young autumn-spawning North Sea herrings drift in as larvae/fry and spend 1.5–2 years in Swedish waters before migrating back and reproducing in the North Sea. They are too small for marketing as consumption fish and the catch is used to produce herring oil and fish meal. Over-wintering adult North Sea herrings have not been found in the Skagerrak since the mid-1960's. The adult herrings caught on the west coast are spring-spawners who spawn during March-April along the coasts of the Kattegat and Skagerrak, in the Belt Sea and perhaps mainly in the southwestern Baltic. After spawning they migrate all the way out to the northeastern part of the North Sea. They return during the autumn and have an important over-wintering area in the Sound.

Mainly spring-spawning herrings are fished in the Baltic. Herrings landed to the north of Kristianopel in Kalmarsund are given the Swedish name of *strömming*. The farther to the north we proceed in the Baltic, the smaller the *strömming* gets. An adult *strömming* in the Bothnian Bay weighs about 50 g, a herring in the southern Baltic of the same age weighs about 150 g, whereas an Atlanto-Scandian herring weighs about 500 g. For centuries, the dominating fish species in Swedish fisheries has been the herring. In 1984 its total catch value was exceeded by cod.

Today, catches of herring are limited more by demand, and thus by price per kilo obtained by the fisherman, than by availability. The Baltic herring has difficulty in competing with the larger North Sea herring on account of its smaller size.

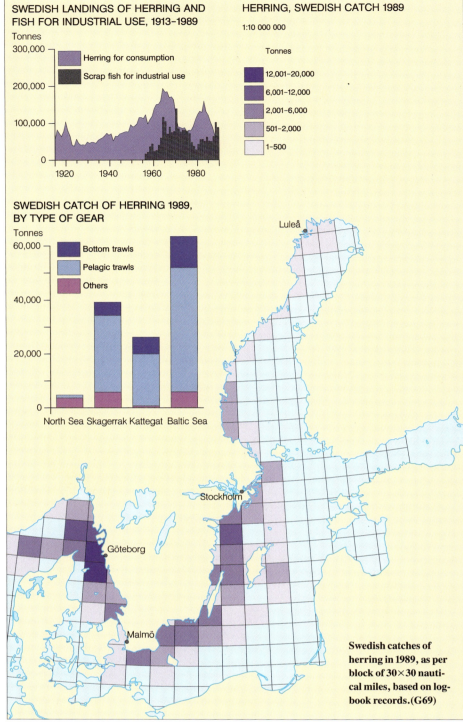

Swedish catches of herring in 1989, as per block of 30×30 nautical miles, based on logbook records.(G69)

Catch areas for shrimp. In 1989 the world catch of shrimp was about 188,000 tonnes.(G70)

Prawns and Shrimps

The deep-sea shrimp (*Pandalus borealis*) belongs to the same group of crustaceans as lobster, crab and Norwag lobster, i.e., decapods.

The deep-sea shrimp may reach a size of more than 15 cm. The shrimp lives at depths of 150–400 m and is also present in some of the deeper fiords on the west coast, such as Gullmarsfjord and Kosterfjord. The deep-sea shrimp has an unusual reproduction strategy, it is a protandrian hermaphrodite, i.e., the same individual functions first as a male and then as a female.

Shrimp eggs are hatched in the spring and the free-swimming larvae live a planktonic life for about three months until they move down to the bottom. At an age of 1.5 years, the shrimps have developed into males and participate in mating during the autumn. During their next year of life, the male organs degenerate and are successively replaced by the female organs. Most shrimps function as females during the autumn when they are 2.5 years old. They remain as females for the rest of their lives. After mating, the females, depending on their size, produce 1,000–3,000 eggs, which are carried between the legs of the abdomen. In the spring, when the eggs are ready to hatch, the females move to slightly shallower waters (about 150 m deep).

The shrimp lives on various kinds of bottom-living animals such as bristle-worms, crustaceans and residues of dead plants and animals (detritus). During the night, when the shrimps move up towards the surface, they catch copepods and krill (*Euphausiacae*).

Swedish shrimp-fishing with bottom-trawls (beam trawls) started in 1902 when a fisheries biologist with a newly-designed trawl was able to demonstrate that vast quantities of shrimps were present in Gullmarsfjord. Fishing started in Gullmarsfjord and Kosterfjord and later spread to several areas in northern Bohuslän. The fishing was done in the archipelagos and the annual catches were about 100 tonnes. During the 1930's the shrimp-fishing fleet expanded as well as the area fished. Annual catches rose to about 1,000 tonnes and shrimp-fishing took place along the southern edge of the Norwegian Channel and up towards Egersund with a fleet of about 150 vessels.

The large increases in catches in the early 1960's were possible through a combination of good year-classes of shrimps, an increase of the size of the fishing-fleet from 200 to about 300 vessels, and radical improvements to the trawls. The crash in population and decrease in catches in 1966 led to many fishermen giving up shrimp-fishing.

In Swedish shrimp-fishing the large individuals are the ones most in demand. The catch is sorted on-board and the largest shrimps (the females) are boiled immediately. The smaller fraction is iced and sold to the canning industry or as bait. The price differences of these categories are considerable: in 1989 boiled shrimps fetched 52 SEK/kg and canning shrimps 14 SEK/kg. The smallest fraction remaining after the grading is thrown overboard, which may represent considerable numbers which are lost in this way.

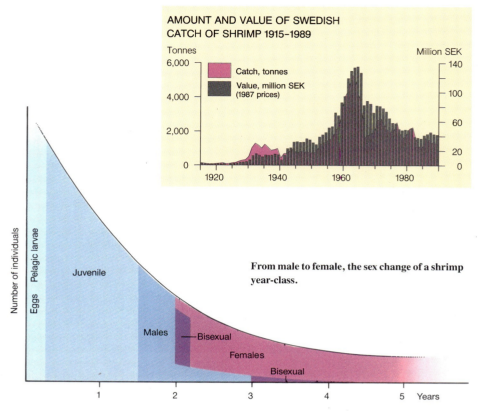

From male to female, the sex change of a shrimp year-class.

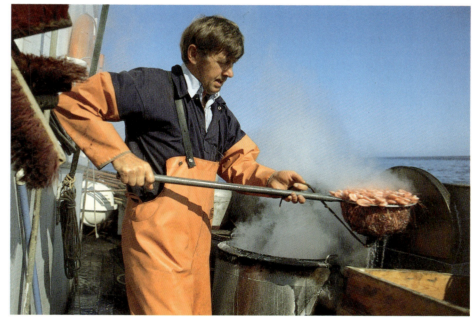

Shrimp-fishing off Smögen. The shrimps are boiled on board immediately after being caught.

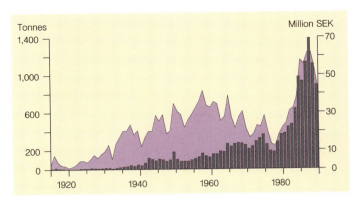
AMOUNT AND VALUE OF SWEDISH CATCH OF NORWAY LOBSTER 1915–1989

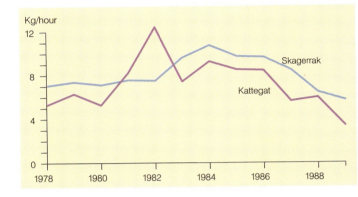
CATCH OF NORWAY LOBSTER PER TRAWLING HOUR

Catch areas for Norway lobster. In 1989 the world catch was about 55,000 tonnes. (G71)

Norway Lobster

The Norway lobster (*Nephrops norvegicus*) is also called the Dublin Bay prawn. It differs from a "real" lobster by being smaller, thinner and with narrower claws. It has a pale red-yellowish colour on the back and yellow-white on the belly. Max. lengths are about 0–24 cm.

In Swedish waters, the Norway lobster is found at depths of ca. 30–250 m. A characteristic of the Norway lobster is its burrowing way of life. Thus, it is restricted to soft bottoms where the mixture of sand and clay is suitable for building galleries and cavities. The galleries reach 20–30 cm into the bottom and may vary from a simple U-shaped gallery to complex systems of tunnels with cul-de-sacs and several exits. The Norway lobster leaves its hole when light penetrating down to the bottom has reached a suitable level. At a certain depth, this might occur at dusk and dawn but may be around midday in greater depths of water. It also changes with season. Larger individuals spend a longer period outside their holes on the seabed than small individuals.

The females spawn every second year and the eggs are carried on the underside of the abdomen for 7–9 months. When newly-laid (in August-October) the eggs have an intense green colour but darken and turn black after about a week.

Norway lobsters are generally fished with special trawls which are rigged so that they penetrate deeper into the bottom than in other fishing. Recently, double-trawls with two cod-ends and a heavy metal weight between them have been introduced. Baited cages in series have been used on the west coast since the mid—1980's. Fishing for Norway lobster has become one of the most economically important branches of Swedish fisheries. No estimates have been made on the population size of Norway lobster in Swedish waters. However, catches per fishing-hour in the Kattegat and Skagerrak have decreased since the mid—1980's, indicating reduced populations.

ENVIRONMENTAL INFLUENCE

The Norway lobster has a well-developed ability to survive in low oxygen concentrations. In the same way as when humans are in "thin" air on high mountains, it can increase the amount of oxygen-absorbing pigment in the blood and thereby, if oxygen decreases slowly, survive at an oxygen concentration that is so low that most other mobile species have moved elsewhere. In the event of abrupt decreases in the oxygen concentration on the bottom, they will emerge and stand "on their toes" in order to get as far away from the bottom as possible.

In the Kattegat, the hydrographic conditions are such that there is a regular decrease in the oxygen concentration in bottom water during July-September. This seasonal decrease in the amount of oxygen has been enhanced during the 1980's as a result of increased supply of nutrients whereby a total oxygen deficiency (<1 ml/l) has occurred on several occasions. The south-eastern parts of the Kattegat (Kullen-Falkenberg) have suffered the most, implying that the population of Norway lobster has been severely decimated and that fishing has ceased.

The Norway lobster lives on soft bottoms, where it builds galleries and cavities.

Catch areas for salmon. In 1989 the world catch of salmon was about 10,200 tonnes. (G72)

Salmon

The salmon (*Salmo salar*) is a north Atlantic fish which lives pelagically in the upper water layers. It migrates over extensive distances; salmon marked in west European waters have been found along the coasts of Greenland as well as in the White Sea. The Baltic salmon mainly remains in the Baltic but occasional finds have been made elsewhere.

Male salmon grow to a length of almost 150 cm and weigh up to 50 kg, the females reach lengths of 120 cm and weigh about 30 kg. Salmon spawn in running water, generally in the upper reaches of rivers. In the spring, the salmon comes to the coasts and the estuaries. At this time it is fat and in good condition. It uses its energy reserves to swim up the river, past waterfalls and rapids. Jumps of 3–4 metres and swimming more or less vertically up a waterfall require great strength. Large amounts of energy are also used to form eggs and sperm. A salmon female is capable of producing eggs corresponding to 25% of her body weight.

The 2 cm large larvae, which have a large yolk sac remain in the bottom gravel for about a month until the yolk sac has been consumed. The young salmon remain in the river for about two or three years. After it has become adjusted to life in salt water, it is called smolt. This adaptation can be seen by the disappearance of the grey-blue spots on the sides and the fish becoming silvery in colour. Grilse are salmon which migrate up the rivers after only one year in the sea. Most of the salmon, however, remain for up to 4 years in the sea before returning to spawn in their native waters. The spent salmon are called kelt after spawning.

Salmon are fished both in the sea, along the coasts when returning to their native waters, and also in the river itself as they migrate up to the spawning localities.

The places where salmon may be caught have long been the subject of international dispute. States with salmon rivers consider that salmon should only be fished when they have returned to their home river. This approach was formerly difficult to accept by the non-salmon producing countries but today it has finally been accepted and salmon fishing in the north Atlantic is banned. In the Baltic, on the other hand, no such regulations have been introduced and most of the catches are taken at sea. The fishing in the open sea is done with drifting salmon nets and with salmon lines, i.e., floating long lines. Fishing along the coast is done with fyke nets and pound nets.

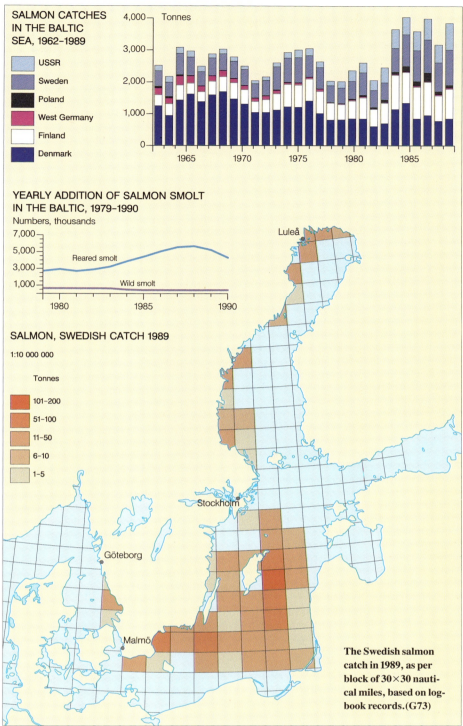

The Swedish salmon catch in 1989, as per block of 30×30 nautical miles, based on log-book records. (G73)

THE SALMON PARADOX

The Baltic salmon is threatened by extinction but today there are more salmon than has been the case for many years. How can this be explained?

The spawning-places of the salmon have been reduced drastically since the 1950's as a result of regulations, hydro-electric installations, polluting substances and barriers of fishing tackle in the river estuaries. The power stations have been ordered by the Water Rights Courts to rear salmon to the smolt stage and Sweden annually releases about two million reared salmon in the Baltic and about the same amount is released in Finland. However, this implies a risk for serious genetic depletion.

Herrings before being smoked. The hot-smoked fish are marketed as *böckling*, or smoked Baltic herring.

Fish Processing

Fish is a perishable food which has poor keeping qualities in comparison with, e.g., meat of terrestrial animals. Since ancient times, attempts have been made to improve the keeping qualities of fish by using different means of conservation. Drying and smoking are the oldest methods.

Drying has a conserving effect since water is removed from the fish and in this way makes it less attractive to bacteria. Ling is the only dried commercial fish product in Sweden today. The ling is cleaned and salted before being stretched on wooden pegs on the skin side and hung up to dry. It is used for stockfish. Domestic drying of cod, flatfish, pike, perch and roach is done locally.

Smoking both preserves and also improves the taste of the fish. Both the heat, the salting as well as bacteria-killing substances in the smoke contribute to improving keeping quality. In cold-smoking the temperature is below 25°C and by means of salting the fish is kept fresh for the relatively long period (up to three days) required by the smoking process. Mainly salmon are cold-smoked.

Hot-smoking is done at about 60°C but is sometimes ended with an increase of the temperature to about 100°C. Eel, herring, powan, mackerel, and rainbow trout belong to the assortment of hot-smoked fish.

PRESERVATION BY SALT

Salting is a preservation method which is almost as old as drying. Many species have been salted but it is mainly herring which has been salted in large amounts. Access to salt has been a condition for benefitting from the great shoals of herring that have occurred ever since herring fishery began in Skåne during medieval times, via the herring boom of Bohuslän to the situation today. Salt herring can be stored for about a year and this enabled it to be transported to the population centres in the cities. During the late 18th century, when herring fishing was at a maximum, the salting and the herring-oil industries flourished in Bohuslän. In 1786/87 a total of 334 industries for salting herrings were registered at the County Offices in Göteborg. Salt herring is still an important raw product for the canning industry.

FREEZING

Low temperature has an inhibiting effect on bacterial development but usually does not kill bacteria. Techniques for freezing and storage of fish are mentioned as early as the 19th century (in a patent granted to E. Piper). Both the technology and the popularity of the method expanded rapidly in the USA during the 1920–30's. Today it is the dominating preservation method in many places, including Sweden. Mostly fillets of fish are frozen but also whole fish.

FISH PRESERVATION

A preserve usually implies "a product which is given a greater or smaller degree of durability by being enclosed in air-tight (hermetically) sealed vessels of metal or glass". The technique was described in 1810 by the French chef Nicolas Appert and rapidly became popular. Appert boiled glass jars containing fruit, vegetables or meat in a salt solution and was thereby able to increase the boiling-point to above 100°C. In England, instead of glass, B. Donkin used jars of tin-coated metal-plate, tins, on which the top and bottom were welded into place. Today the market is dominated by aluminium cans with a top which is seamed into place. Hermetical preservatives are either fully- or semi-sterilized tinned goods. Fully-sterilized tinned goods are sterile as a result of being heated to at least 110°C in an autoclave and have, in principle, unlimited durability. Examples are sardines fish-balls, and mussels.

Semi-sterilized tinned goods, on the other hand, have limited keeping quality, which may be prolonged if they are kept chilled. They are not sterile but the activity of the bacteria has been inhibited by additions of salt, sugar and/or vinegar. Examples of semi-sterilized goods are: Anchovies, pickled herrings and caviar. A special form of semi-sterilized goods is the fermented Baltic herring (*surströmming*) prepared in Norrland. It is preserved by means of microbes: lightly-salted herring is allowed to ferment anaerobically. The process is carefully controlled and terminated when the correct taste has been achieved. This method of preserving fish is very old and has concerned species such as roach and trout. Fermented fish is characterized by a pungent odour in which methylmerkaptane is an important ingredient. Few people remain unmoved by this odour: many, not only inhabitants of Norrland, have learnt to appreciate the fish and consider it a delicacy, whereas others consider that "Fermented herring is a salt and rotting herring which has a sour, unpleasant taste and an undescribably revolting smell".

The first Swedish canning factories were opened in Fjällbacka, Lysekil, Marstrand and Gullholmen in the mid—19th century. These factories canned sprats cured with salt, sugar and spices and called them *ansjovis* – anchovies. This method of preserving sprats had been known at least since the mid—18th century and Scandinavian anchovies were in demand even abroad. The sprats (*Sprattus sprattus*) were available in abundance whereas the herring had disappeared from the west coast by the early 19th century.

The booming sea transports with sailing vessels offered good distribution opportunities and the demand for anchovies was good not only in Sweden but also in other countries. In addition to the anchovies, the factories gradually started to produce caviar, mackerel products and sliced fillets of Atlanto-Scandian herring.

After the First World War the can-

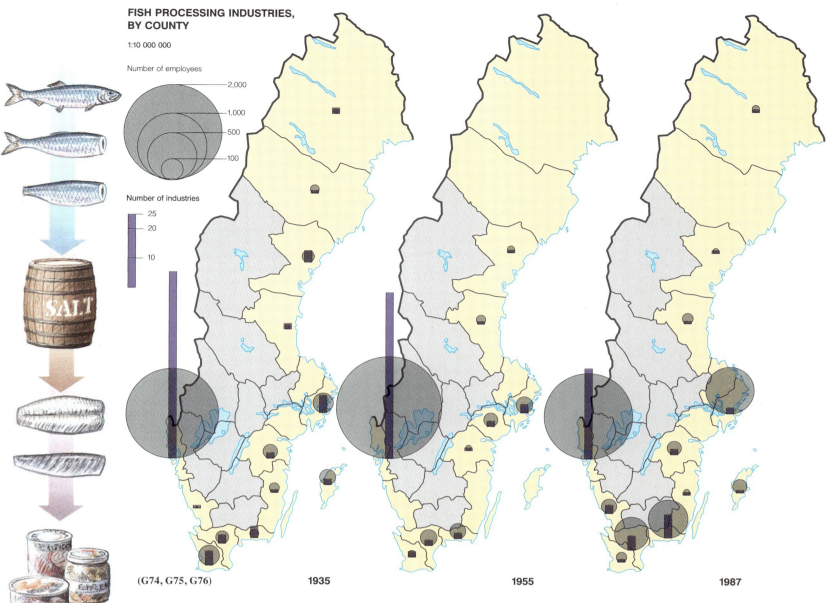

FISH PROCESSING INDUSTRIES, BY COUNTY
1:10 000 000

Number of employees: 2,000 / 1,000 / 500 / 100

Number of industries: 25 / 20 / 10

(G74, G75, G76)   1935   1955   1987

Interior of a fish factory. Cleaning and filletting fish is traditionally done by women.

ning industry expanded steadily: the number of factories increased from about 30 in 1920 (with about 500 employees) to about 90 in the 1950's (about 3,000 employees). The range of products also increased. The sliced fillets of herring soon became the most popular product. Christmas anchovies were introduced and soon became an essential part of the Swedish Christmas table. Sprats prepared as sardines became a major product in the 1930's and fully-sterilized products of crustaceans such as shrimps, crayfish and mussels were introduced.

When the availability of large herring decreased (Atlanto-Scandian herring, North Sea herring) and was replaced by smaller herrings from, e.g., the Baltic, this also had repercussions on the range of herring products: The marinated herrings in glass jars increased at the expense of the sliced fillets and other products which were based on salted herring.

In the 1970's there was a marked structural reorganization within the canning industry: a decreasing number of companies, decreasing numbers of employees, closures on the west coast, re-establishment on the east coast. Among the important factors in this respect can be mentioned the changed availability of raw products following the collapse of the Atlanto-Scandian herring populations and as a result of Norway and the EC extending their fishing limits, as well as changed consumer patterns resulting from reduced salaries in real terms, and subsidies introduced to support beef, pork and cheese production (in 1973). Other factors were the changed competitive conditions; the free trade convention with the EC in 1973 did not include prepared and preserved fish products which resulted in tariffs (up to 20%) when exporting to EC countries. There were also changes in fisheries policy which implied that the dominance of herring in Swedish fisheries up to that time was decreased in favour of increased cod-fishing. The fisheries also became concentrated more in the Baltic.

The range of products from the canning industry today has increasing shares of frozen fish and ready-cooked products such as fish sticks and baked fish. Modern consumers appear to prefer something which can be rapidly heated in the microwave oven instead of having to clean and prepare a fresh fish or – heaven forbid – having to lift a salted herring out of a keg!

FISHING OCCASIONS AND TYPICAL CATCH SPECIES

1:10 000 000

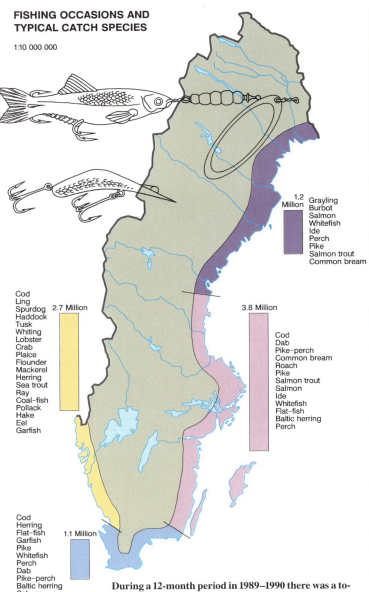

During a 12-month period in 1989–1990 there was a total of about 26 million fishing-occasions in all Swedish waters, of which one-third (9 million) in the sea. (G77)

Angling is the commonest form of sport-fishing. Here at Stenshuvud, to the south of Kivik.

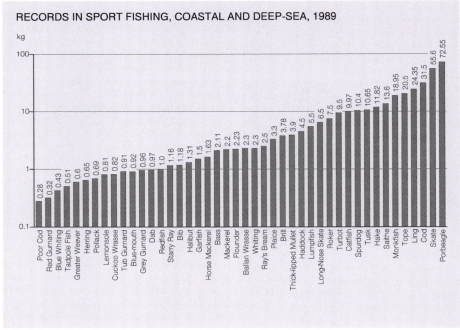

Leisure-time Fishing

Leisure-time fishing is done both as recreation and to obtain fish for personal use. When emphasis is placed on obtaining fish for the "kitchen" and for one's own freezer we speak of domestic fishing. The tackle used may range from nets, fyke nets, cages and seine nets to jigs, hand-lines and fishing-rods. In principle, all types of tackle except for trawls. Since 1986, the use of trawls has been restricted by law to professional fishermen.

To be called a sport fisherman you must restrict your use of tackle to hand-tackle, e.g., rods or activities such as spinning behind a row-boat, harling, trolling, jigging, dibbing or fishing with a hand-line.

Fishing for home consumption is an ancient activity and has been a complement to hunting and agriculture. It has also been surrounded by rules and regulations. The riparian owner's sole right to fishing has been the dominating principle along the Swedish Baltic coast, as well as in fresh waters. On the west coast this principle has long been restricted in favour of a right for everybody to fish. At present, the riparian privileges on the west coast are restricted to oysters. Since 1970 fishing with both hand-tackle and movable tackle, e.g., nets, has been permitted along the coast of Norrland, but not for salmon. In 1985, fishing with hand-tackle in private waters was also permitted along the east coast.

Sport-fishing in the sea is a relatively new activity. The first sport-fisherman appeared on the west coast in the early 1930's but it was not until after the end of the Second World War that development speeded up. A large inquiry conducted by Statistics Sweden into the fishing habits of the Swedish population, made at the request of the National Board of Fisheries, revealed there were about 2.2 million Swedes in 1990 between 18 and 74 years who had, on at least one occasion during the previous twelve months, done some leisure-time fishing. An estimation of the number of fishing-occasions during the same period suggests that there was a total of about 26 million fishing-occasions, thus more than 10 times per angler. The importance of leisure-time fishing in the sea is clear since 631,000 anglers reported that all their catch had been taken in salt water, 304,000 had taken parts of their catch in salt water whereas 1 037,000 had not caught any part of their catch in salt water. Among those who had caught salt-water fish, 404,000 had used only hand-tackle whereas 61,000 had not used hand-tackle. The inquiry also asked about the size of the catch for different groups of fish. The results suggest that the share of the catch in leisure-time fishing could be about 20% of the total catch in professional fishery (which amounts to ca. 250,000 tonnes) but varies widely from species to species. The share is probably particularly large for expensive species such as lobster and trout but also for pike, perch and pike-

A jigging competition. Jigging is usually done in lakes but also in the sea, mainly along the Baltic coast.

The picture on the top shows trolling.

The table accounts for annual catches of less than 100 kg per species. The total catch for all leisure-time fishermen was 34,100–52,700 tonnes.

RECREATIONAL FISHING-CATCHES 1990

	Tonnes/year
Salmon (Salmon trout, Grayling, Char)	4,100– 5,500
Put-and-take fish (Rainbow trout, etc.)	1,200– 2,200
Carp fish (Roach, Common bream, etc.)	2,100– 3,300
Pike, Perch, Pike-perch	11,300–13,900
Sea fish (Cod, Whiting, Mackerel etc.)	6,100– 9,900
Crayfish	600– 1,600
Other fish species	1,300– 2,500
Total	26,700–38,900

perch, for which Statistics Sweden reported about 1 377 tonnes caught in 1989 by professional fishermen.

SPORT-FISHING

The most common way of fishing in sport-fishing is to use a rod with a casting reel or a spinning reel. Rods with lengths of between 4 and 11 feet long can be chosen depending on the species to be caught and with regard to wind, current and depth. Lines are between 0.07 and 0.55 mm thick and lures may weigh anything between 1 and 150 grammes. The lures may be spinners, spoon baits or wobblers and can be found in a vast variety of shapes and colours. The jigger, who often fishes from the ice, uses a short rod with a simple jig-reel.

Today, trolling is replacing the former habits of putting on home-made spoon baits, holding the line between the teeth and rowing for pike, or spinning for mackerel. Admittedly, a line and a hook is towed behind a boat but the trolling fisherman usually also has other sophisticated equipment. Using an echo-sounder, he can find out the depths at which the fish are present and with a downrigger he can be sure that the hook is at the correct depth.

Sea-fishing from a boat (at anchor or drifting) gives access to species which cannot be caught from the shore, e.g., shark, skate, catfish or ling. Equipment of larger dimensions is needed for this type of sea-angling and jigging since fish of 20–30 kg are quite feasible.

Garfish offer exciting fishing during May-June when they come into shallow waters for spawning. The piers along the Sound between Sweden and Denmark are generally crowded when the garfish come in.

The sport-fisherman considers his catch not only as food but also as a trophy. The aim is to obtain the largest individual of a species. Giant fish are recorded not only in Sweden but also by the International Game Fish Association, IGFA. A large number of competitions, "match fishing", are also arranged where the aim is to catch as many fish as possible in a certain time. During a normal season in Sweden, the sport-fisherman can participate in Swedish and Nordic championships in jigging, Swedish championships in modern and traditional hand-line fishing, as well as championships in ocean-fishing and fly-fishing.

Aquaculture

Aquaculture has long been of importance in food production, particularly in densely-populated areas with moderate access to water. This applies, for example, to China where fish-farming has been well-developed for several thousand years. In sparsely-populated areas with good access to water containing fishable populations of crustaceans and fish, e.g., in the Nordic countries, aquaculture has developed only fairly recently and mainly to produce desirable (expensive) species without any direct importance for the food supply.

In 1988, the world production of aquaculture was about 14.4 million tonnes made up of algae, molluscs (mainly mussels and oysters), crustaceans and fish. This may be compared with the total fish catch which was 98.3 million tonnes in the same year, of which however about 69 million tonnes were used for direct human consumption and which is thus directly comparable with production in aquaculture.

In Sweden, about 6,800 tonnes of rainbow trout and 400 tonnes of salmon in cages were produced in 1988, of which 60% in the sea. The yield of blue mussel was only 860 tonnes, in contrast to 2,560 tonnes in 1987. In addition, 230 tonnes of eels were produced, 100 tonnes of which came from a unit with flowing heated sea-water, and the rest from units with recirculating fresh water. In fresh water, there were also 80 tonnes of char, three tonnes of crayfish and 3.5 million smolt of salmon and sea trout for compensation releases.

Many different methods are used in aquaculture: in extensive farms the organisms are kept in relatively sparse populations and are completely or partly dependent on food which is produced naturally in the same water. In farms with intensive production, smaller units are used, generally ponds, with dense populations and well controlled water environment. The feeding is adapted to the maximum growth capacity of the fish and the feed is frequently artificial dry feed. In this type of production we find units with recirculating water in closed systems, high-tech production units with advanced equipment for heat-pumps and heat-exchange, water purification and oxygenation as well as many other different types of pond enterprises. Examples of extensive methods are blue mussel farms, certain types of pond and lagoon production and "ranching". Ranching implies that young individuals are released into the open water, generally after a short stage of intensive production, and the fish find natural food for themselves. This production method was first used for salmonids on the American west coast. Here, the ability of the salmon to find its native waters when it returns on its spawning migration is utilised. Even an artificial current in the sea is sufficient to awaken this instinct in salmon if the smolt have been released there. Other species have later been used and even mussels and snails have been placed out on suitable seabeds when such populations have not been present for various reasons. The harvesting also takes place at or close to the points of release.

Modern fish-farming started with the production for compensation release and was developed in Sweden during the 1940's. The hydro-electric industry has been obliged by law to compensate for damage caused to fish populations when rivers were utilized for production of electricity. This compensation was done by release of artificially-reared salmon. The principles for pool production using dry feed have subsequently been adapted to many different species.

Production of table fish in cages in coastal waters started to develop during the 1960's, where Norway has been at the forefront on account of favourable conditions both as regards water temperature and sheltered positions. Mainly salmon and rainbow trout are produced in temperature-controlled water but other species, e.g., sea char are starting to become popular.

In production of blue mussels, use is made of nature's surplus production of mussel larvae. When the larvae leave their pelagic stage and settle onto a substrate, ropes or bands are suspended in the water which are soon filled with small mussels. The mussels filter off their feed from the water, mainly unicellular algae, from the water and consequently there are no feed costs. Unfortunately, some algae produce poisons which do not influence the mussels but are accumulated in them. If they are subsequently eaten by humans, the poison may cause diarrhoea and occasionally even death. Farm-produced mussels are examined before harvesting and if poisons are found, then a retailing ban is imposed. This is one of the reasons for the low Swedish production of blue mussels in 1988.

In intensive production, the animals are fed with the maximum amount of feed in relation to temperature and other conditions. Fish can utilize about half of the food given, whereas the rest enters the environment in different forms. A large unit producing several hundred tonnes of fish per year will have a strong influence on the water environment, at least locally. The total Norwegian salmon production in cages at the coast (120,000 tonnes in 1989) results in a discharge of nitrogen and phosphorus which is equivalent to the amount released by the entire Norwegian population.
A transfer to land-based production where the water can be filtered and purified in other ways before it is returned to the sea, lake or river from which it was taken, is therefore desirable.

Rearing of salmon at Bergforsen. The smolt are released into the Baltic Sea to compensate for destroyed spawning grounds.

Cage production of fish. This type of production is not particularly popular in Sweden owing to water temperature and ice conditions.

Production of blue mussels is done by means of ropes suspended from buoys. The mussels settle on the ropes or bands and obtain their nutrients directly from material occurring naturally in the water.

In Sweden, there has long been control of fish diseases in aquaculture, which is today organized by Fiskhälsan AB (Fish Health Co.). The relatively thorough control and the strong restrictions applying to imports of live fish have succeeded in keeping Swedish aquaculture free of most of the diseases which are difficult to treat. However, the problem has increased during recent years partly since some producers have been using inexpensive stocks with poorly controlled quality in order to reduce production costs. Transport of fish between different units has increased for the same reason and a certain amount of illegal import of fish has clearly taken place and has resulted in the arrival of new diseases.

Increasing production costs during recent years have not been compensated for by increased prices and thus certain species have probably reached a peak in demand and thus in production volume. This applies, for example, to rainbow trout and salmon; several production units closing during 1989 both in Sweden and Norway. Consequently, for these species there are probably only marginal developmental opportunities. For some species, mainly marine species such as turbot and cod, the cultivation biology and technology is very complicated and has not been established to any particular extent until recently. Here, however, there are potential developmental opportunities since commercial activities have been started and method developments may be expected to reduce production costs considerably.

Shipping and Navigation

Boating is an important form of relaxation. Smögen, in Bohuslän.

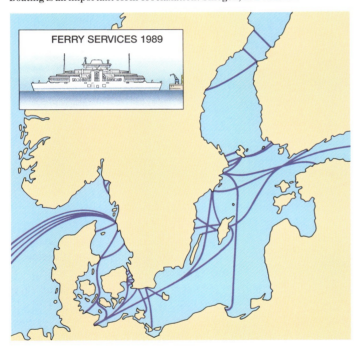

Ferry routes between Sweden and neighbouring countries. (G70)

Shipping and navigation, which closely involves shipping companies, ports and ship-building, is usually divided into different kinds of shippping; merchant shipping, fisheries and recreational boating.

The development of world shipping is strongly influenced by fluctuations in the price of oil and the exchange rate of the dollar. It is commonly believed that global shipping will develop strongly during the 1990's. Developments of this kind will naturally also influence shipping, ports and shipping companies in Sweden.

When it comes to transporting large quantities of goods at low cost, there is still nothing to compete with merchant shipping.

Today there are about 250 ports in Sweden, of which 51 are public ports responsible for most of the country's turnover of goods. Technical developments in load-handling technology have implied major demands on re-organization in the ports. The introduction of containers and ro-ro vessels (roll on, roll off) has led to the ports needing corresponding changes in their activities. This has resulted in many older ports, which were unable to keep up with the developments, being in poor condition today and having to face the choice of either acquiring better equipment or closing down. These are questions of major importance for future developments in several coastal communities.

The port structure of the future will also have internationalization as prime mover, i.e., international development in this sector will have a controlling effect on developments in Swedish ports. A land connection over the Sound will, in this context, have consequences for the transportation branch throughout Sweden. As regards shipping, the effects will largely depend on how this link is financed. In addition, ports will be influenced by the development of domestic coastal shipping in relation to transport by truck or rail.

The importance of the smaller ports as feed-ports and terminals may increase in the future as a result of discussion on the environmental effects caused by the various transport systems. In this context, coastal shipping implies a lower environmental load when different goods are to be transported over long distances. With the improvements already made to reduce the environmental effects of shipping, together with those already planned, the long-term perspective of shipping today remains a means of transport with good competitive ability from the environmental viewpoint.

The development of ferry traffic is also linked with international developments in shipping, not least as a result of developments in cars and tourism. When measured in volume, ferry traffic deals with about 20% of all transported goods, but as regards value, the corresponding figure is about 50%. Ferry traffic is thus responsible for the transportation of more expensive products. Changes

Container freighters being loaded in the port of Göteborg.

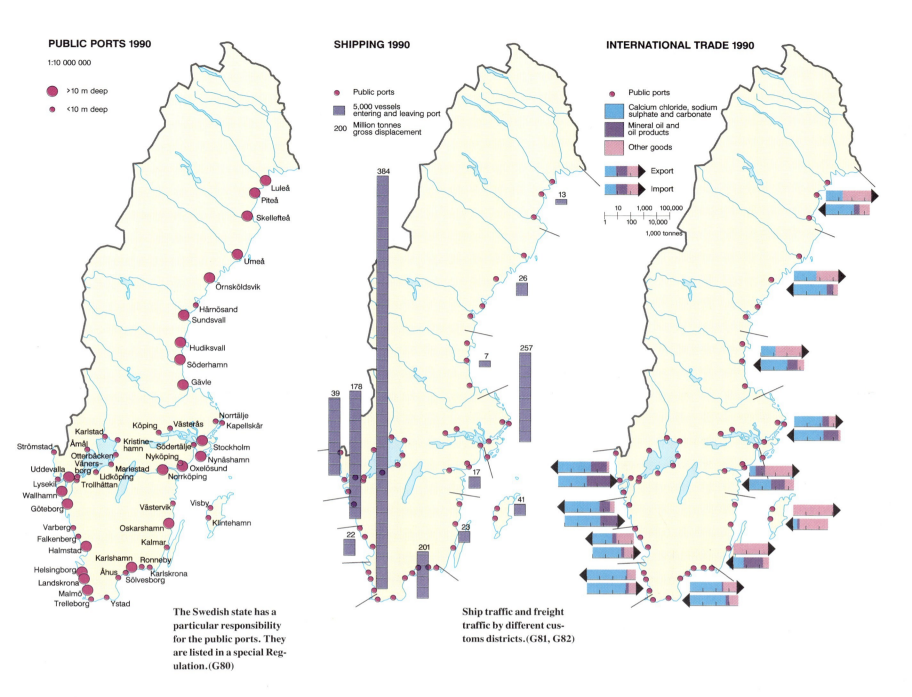

PUBLIC PORTS 1990

The Swedish state has a particular responsibility for the public ports. They are listed in a special Regulation. (G80)

SHIPPING 1990

Ship traffic and freight traffic by different customs districts. (G81, G82)

INTERNATIONAL TRADE 1990

taking place in east European countries are creating opportunities for more dynamic commerce and traffic within the Baltic area, an area particularly suited to ferry traffic as a result of the short distances.

Recreational boating has increased strongly during the past ten years. This is noticeable not only in the increasing number of recreational boats of different kinds but also in the demands for better port facilities and channels used by recreational boats. This is a new type of demand and is influencing the life and development of coastal communities. The most attractive areas are coastal stretches with broken coastline and archipelagos and here the development of recreational boating has been most obvious. The demands placed have sometimes led to conflicts with already established interests, especially with regard to speed restrictions and choice of locations for marinas, etc.

Environmental Effects

Environmental effects of shipping are associated with emissions of different kinds to water and atmosphere. These emissions occur in connection with both the normal operation and maintenance of vessels as well as in port activities and accidents.

Earlier, water pollution caused by vessels was mainly noticed in connection with the handling of rubbish, oil and solvents, together with the emissions of oil and chemicals resulting from shipwrecks or other accidents. During recent years, air pollution and other influences caused by shipping have attracted increased interest, particularly discharges of exhaust gases from engines. In some places, these may have an important environmental effect on air quality, such as in Stockholm, where air pollution caused by ferry traffic makes up about 20% of the emissions of nitrogen oxide. Other effects on the environment are caused by noise, vibrations and shoreline erosion by waves, etc.

MARPOL 73/78 is an international convention for protection against pollution from vessels. According to the convention, the Baltic is classed as a particularly sensitive marine area. The convention, which covers the entire world, is applied by the Baltic States through the Helsinki Convention. Environmental problems involving shipping and navigation are dealt with by a special committee. The work of the committee and the measures taken by the individual Baltic States and shipping companies have led to the number of oil emissions in the Baltic decreasing during recent years. In Sweden, for example, the number of accidents with oil emissions has decreased from about 50 per year during the 1970's to 15–25 per year during the 1980's.

Energy

A platform for oil and gas prospecting in the southern Baltic Sea.

The sea contains a number of different potential energy resources which can be extracted in different ways from the bedrock beneath the seabed, from the movements in the water and from the wind above the surface of the water. In this context the exploitation of oil and gas probably comes first to mind, despite these not being specifically marine sources of energy. In the case of Sweden, there are interesting prospecting areas to the east and south of Gotland, in Hanöbukten and to the south of Skåne. Apart from oil and gas resources beneath the floor of the sea, there are genuine marine energy resources which can be obtained from the water itself, e.g., extraction of energy from tidal currents and waves. Today, power stations of this kind are only in the experimental stage. In addition, sea-based wind-power may also be considered to be a marine form of energy.

Among the alternative "marine" forms of energy, it is mainly wind energy and to some extent wave energy which are today of interest in Swedish marine areas. The decisive question for their future commercial use is whether technical developments in wind-power and, in the longer term, wave-power, will make them sufficiently competitive in the national economy with regard to alternative methods of producing electricity.

The purely physical conditions for using the different forms of energy vary between different areas of sea. Developmental work is ongoing for some forms of energy. In Sweden, sea-based wind-power has proceeded furthest in its development. Both physical and technical conditions are such that wind-power is considered to have strong possibilities of becoming an important source of energy in the future. Experimental activities with individual units at places along the coast are in progress. In the future, when wind-power is believed to be of greater importance, it is planned to build complete wind-power stations on floating caissons on an industrial scale before towing them out to suitable sites where they will be sunk onto prepared foundations. The individual wind-power units will then be linked together into a group station which may comprise up to about 100 wind-mills.

Wave energy technology, using different technical systems, is also been developed. So far it is mainly small-scale technology for providing electricity for beacons and measuring buoys that has proceeded farthest in the development. In one approach, the large-scale wave energy technique is based on large buoys linked together to form a wave-power station. The wave climate along Swedish coasts offers good opportunities for this type of energy production. However, no further developments of large-scale wave energy techniques are in progress in Sweden today.

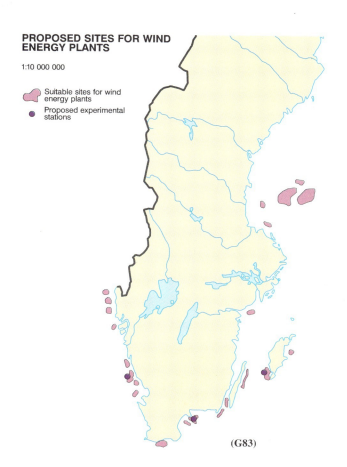

PROPOSED SITES FOR WIND ENERGY PLANTS

1:10 000 000

- Suitable sites for wind energy plants
- Proposed experimental stations

(G83)

Wave energy converters being placed in position for testing at Vinga.

The world's first sea-based wind-power plant, SVANTE. The power plant is an experimental unit located off the coast of Blekinge with an effective output of 220 kW.

117

Emissions

Many polluting substances are transported out to sea in river water. Idefjorden, 1973.

The rich growth of green algae on stones and jetties is the first sign that the water is over-fertilized. Norrpadda archipelago, Stockholm.

In discussions on marine resources, it is sometimes considered that the capacity of the sea to receive pollutants is a resource. Regardless of our attitude to such an approach, the sea is, in fact, utilized in this respect and has been exploited by industries and communities in coastal areas for many years. One of the problems has been that we do not know how much and what the sea can accept, another is that discharges take place in areas which are important and sensitive from biological-ecological viewpoints. The restrictions placed on discharges by regulations have been insufficient and the negative effects have become increasingly problematic and controversial.

The problem of marine pollution is not just a question of direct discharges into coastal waters from industries and communities but also on indirect emissions through runoff from land and atmospheric deposition. Today, the supply of pollutants from rivers and the atmosphere is the largest contribution to the pollution load on Swedish marine areas. By improving municipal sewage treatment plants and corresponding plants in industry, direct discharges have been reduced considerably during recent years.

As regards the magnitude of emissions of pollutants to Swedish marine areas, it is impossible to study only one country. About 25 million people live in coastal communities around the Baltic and about 80 million people within the entire drainage area. Even if restrictions on emissions in Sweden may lead to local improvements in the environmental situation, the condition in the open sea depends entirely on a reduction in the load emitted from point sources as well as non-point sources in the drainage area. The environmental situation in the Baltic and measures to improve it are therefore a sector of international cooperation requiring the highest priority by all coastal states.

Nutrients

Since the turn of the century the supply of nitrogen and phosphorus to marine areas around Sweden has increased many times over. The fastest increase was during the 1960's and 1970's. The use of fertilizers in agriculture and modern forestry has been the main contributor to this situation. The increasing nutrient load also depends on increases in the deposits of nitrogen from the atmosphere and in precipitation. A large share of this nitrogen originates from combustion, in which increased road traffic plays a major role.

The upper table shows the direct discharges of phosphorus and nitrogen from Sweden and the percentage share of the total discharges to the different areas. The lower table shows the supply of metals by atmospheric deposition and emissions from land.

PHOSPHORUS AND NITROGEN SUPPLY

Drainage area	Phosphorus tonnes/year	%	Nitrogen tonnes/year	%
Bothnian Bay	1,000	29	19,000	28
Bothnian Sea	1,600	37	35,500	30
Baltic Proper	1,780	5	44,300	6
Kattegat	900	29	37,000	46

EXTERNAL SUPPLY OF METALS, TONNES/YEAR

	Hg	Cd	As	Zn	Cu	Pb	Ni	Cr
Bothnian Bay	0.5	3	13	554	120	60	28	12
Bothnian Sea	1.3	7.5	13	1,500	245	180	42	70
Baltic Proper *	8	62	60	4,500	1,330	770	81	35
Kattegat and the Sound	1.5	5	15	450	80	108	25	9

* incl. Gulf of Finland and Gulf of Riga

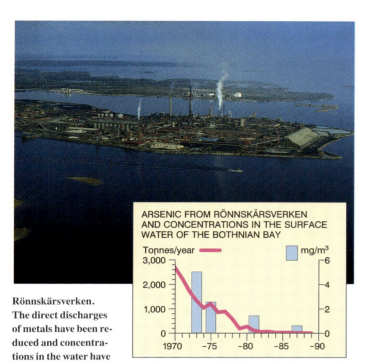

Rönnskärsverken. The direct discharges of metals have been reduced and concentrations in the water have decreased.

ARSENIC FROM RÖNNSKÄRSVERKEN AND CONCENTRATIONS IN THE SURFACE WATER OF THE BOTHNIAN BAY

SUPPLY OF PETROLEUM HYDROCARBONS TO SWEDISH COASTAL WATERS

Source	Tonnes/year
Municipal outlets	16–800
Refineries	70
Other industry	70
Surface runoff	200
Rivers	2,700–4,500
Deposition	1,000–10,000
Shipping	900
Total supply	5,000–16,000

REDUCTION OF METAL EMISSIONS
1:10 000 000

Mercury – Hg
Cadmium – Cd
Arsenic – As
Zinc – Zn
Copper – Cu
Lead – Pb
Nickel – Ni
Chromium – Cr

The map shows how much the emissions of metals must be reduced in order to reach a discharge level which is less than 1.5 times the natural supply. (G84)

Metals

Emissions of metals from sources in Sweden have decreased since the 1960's, mainly depending on a reduction of direct emissions from industries. Thus, concentrations in water of arsenic and mercury, for example, have decreased during the 1980's. Nonetheless, emissions of metals from industry and the community are still often several times larger than natural levels. Discharges of metals along Swedish coasts may be considered to be under control today. Continued measures to restrict direct atmospheric and water discharges will further reduce the metal load.

Stable Organic Environmental Poisons

Stable organic environmental poisons are today one of the most serious threats to the marine environment. They consist of a number of different groups of substances, frequently of a mixed character. Today, stable and chlorinated organic substances are sometimes grouped together. This is not fully correct since also certain oil components belong to this group of environmental poisons. Only a couple of per cent of the stable organic substances are known. They have been spread both knowingly and unknowingly in the environment and today can be recovered from the marine flora and fauna, in some cases with extremely high concentrations in fish and mammals. Bleaching plants of pulp industries are the dominating source of emissions of chlorinated organic substances in Sweden and, in pace with improved knowledge on these substances and their effects on the environment, there may well be several unpleasant environmental surprises awaiting discovery in the future.

Emissions of oil in the marine environment are a serious problem. Direct emissions from, e.g., shipwrecks, result in acute local pollution and cause evident effects such as mass mortality of seabirds. The effects of indirect emissions of oil pollutants in sewage water, runoff from land, and atmospheric deposition are not so noticeable despite them making up more than 90% of the supply. There is still major uncertainty as to the sources of indirect, diffuse oil pollutants and their effects on the environment. Major concern has been expressed that both direct and indirect emissions of oil will increase in the future in connection with the development of the off-shore industry in Swedish marine areas.

Mineral Resources

Economically the most interesting marine deposits of ore and minerals are found within the continental shelf areas of the earth. Ores are found in the underlying bedrock. In shallow marine areas and in estuaries, there may be so-called placers which have been formed as a result of the reworking and sorting by waves and currents of loose sediments into different density fractions. In this way, valuable heavy minerals can become enriched in economically extractable quantities.

IRON-MANGANESE CONCRETIONS
1:10 000 000

- Sporadic occurrence
- General to rich occurrence
- Rich occurrence of manganese-rich nodules

(G85)

Ores and Minerals

The occurrence of industrially interesting ores, minerals and rocks on the Swedish continental shelf is little known. Assumed deposits are therefore based on known deposits from neighbouring land areas and their assumed continuation under the seabed. The position of the Swedish coast in the Bothnian Sea and Bothnian Bay is largely conditioned by zones of tectonic weakness in the bedrock. At some places on land along these zones, titanium, vanadium, intrusions of carbonatites and kimberlites and rocks containing certain rare earth metals have been observed. Between the Sound and the Kattegat there are, e.g., thin layers of coal, ceramic clays, glass sand and limestone within the sedimentary rocks of the Höganäsfält area.

Concretions and Nodules

The lumps of marsh ore found in the lakes also have their equivalents on the continental shelf. These concretions of metals are formed by the organic degradation processes releasing the metals into the pore water of the sediment. Biogeochemical reactions, which are still not fully explained, then lead to the metals in the sediment being precipitated in the form of nodules or as a crust on the surface of the sediment. Otherwise, the nodules are mainly found in the deep ocean areas. Concretions occurring in the Baltic are, apart from the lower metal concentrations, largely similar to those in the deep oceans. They contain about 20% iron and 10–15% manganese together with smaller amounts of copper, nickel, chromium, zinc, cobalt, molybdenon and phosphorus. On the continental shelf they are mainly formed on periodically eroded bottoms, i.e., along slopes down to the deeper basins. Their growth rate (0.02–0.05 mm/year) in such places is greater than in the deep oceans.

Some areas in the Bothnian Bay and Bothnian Sea may be more or less covered by concretions (7–8 kg per m^2). No commercial exploitation is taking place within the Swedish area of the continental shelf.

Sand and Gravel

Perhaps the most important known mineral resources on the Swedish continental shelf are the sand and gravel deposits. At present, these are also the only non-living natural resources exploited commercially in Swedish waters. Exploitation is on a small scale and today is concentrated in certain areas in the Kattegat, the Sound and to the east of Fårö. The marine sand extraction amounts to about 70,000 m^3 or about 100,000 tonnes per year, i.e., hardly 1% of the total extraction in Sweden in this sector. Sand and gravel of poorer quality are used as filling material whereas deposits of high quality (e.g., with high silica contents and low iron contents) are used for manufacturing of cement, glass and glass fibre and within the ceramics industry.

Oil

At present there is no production of oil or gas in Swedish marine areas, but prospecting is being conducted. Between 1969 and 1987 a partly state-owned company has had exclusive prospecting rights on most of the Swedish continental shelf to the south of latitude 59°. Today, other companies, both Swedish and foreign, have been granted prospecting rights. The areas to the east of Gotland appear promising, among others. Since oil production started on the island in 1972 and up to 1989 a total of almost 30,000 barrels (about 86,000 m^3) of low-sulphur oil had been produced. It is mainly the carbonate mounds (piles of calcium sludge fixed by algae) in the sedimentary sequence which have been found to contain oil-bearing structures. However, these structures are often small and large quantities of oil cannot be expected.

In the sea to the south of Öland, there is a narrow north-south zone of alum shale overlaid by sediments. There are also corresponding deposits on land. It has been calculated that the shale contains about 800 million tonnes of oil. During the Second World War, oil was produced on the Swedish mainland from about 50 million tonnes of shale. The richest shale yielded 50–60 litres of oil per tonne.

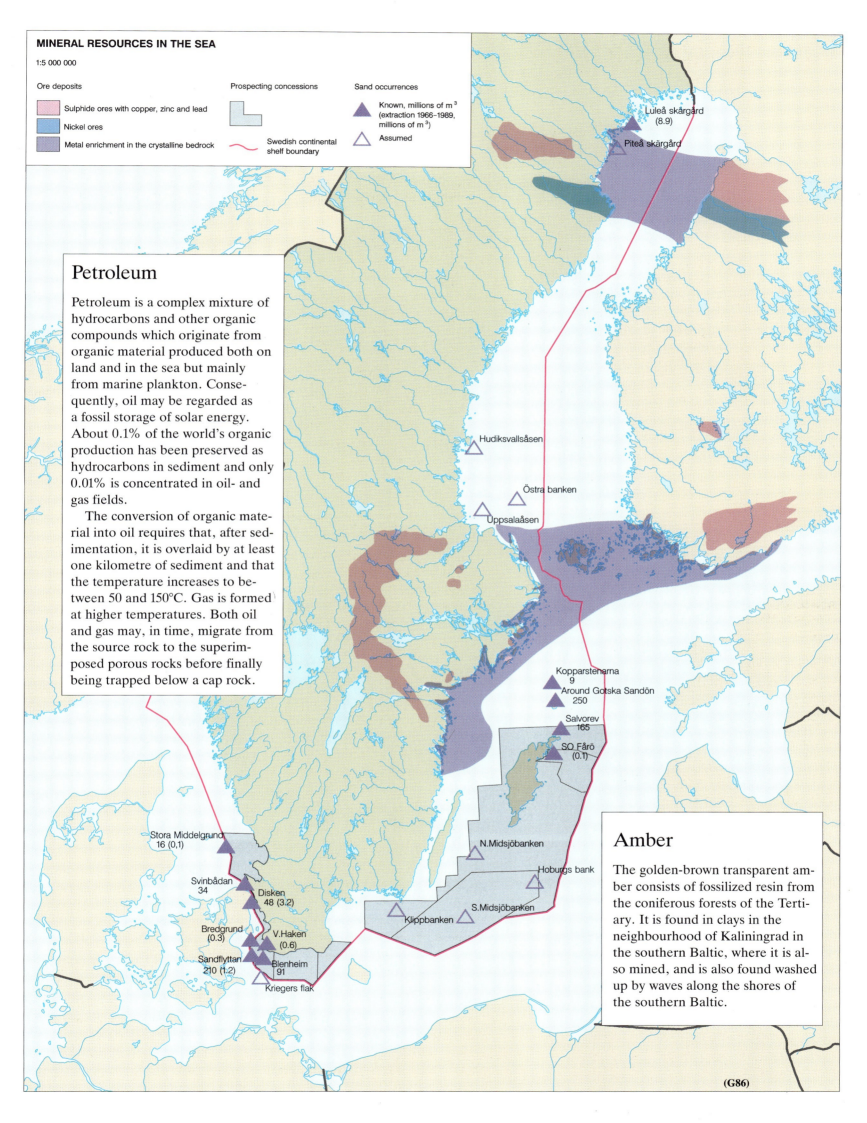

The Changing Sea

Beer cans, milk cartons, oil . . . The shoreline provides good evidence of what coasts and seas are exposed to. The slow changes we can achieve by influencing the global environment are more difficult to discern. We can only guess at what will happen if the water level rises and erosion increases.

Conditions in the sea vary continuously. Particularly large changes are associated with large-scale alterations to the volume of the continental ice (glaciations on a time-scale of 100,000 years). In marine areas close to the large inland ice sheets, the glaciations may have caused particularly massive changes. During the latest glaciation, large parts of the North Sea were dry and the Baltic Basin was filled with inland ice. Since then, as a result of the melting of the inland ice, the surface of the sea has risen by 120 m. When the melting was at its peak, the level of the sea surface rose by about 3 cm per year.

It has generally been assumed that conditions in the open oceans would remain constant in our era. Consequently, it was a major surprise in the early 1980's when it was found that salinity and temperature had decreased in deep parts of the north Atlantic – between 50°N and up to Iceland-Faeroes – since the late 1960's. The decrease in deep-water salinity could be related to an even greater reduction in the salinity of the surface waters in the same areas as well as in areas further to the north. These changes were considered to have been caused by a change in the wind field which caused water with lower salinity to be transported to a greater extent than normal out into the open ocean from the coasts of Greenland and Labrador.

Mankind Influences the Sea

Man has developed from being a sparsely occurring, ecologically well-adapted organism, into a mass-occurring organism, exploiting natural resources to a major degree and actively (both intentionally and unintentionally) interfering with biological development. Today, the chemical composition of water, air and soils is being influenced by man at an accelerating rate, even on a global scale. Within a fairly short perspective, this may also become of importance for the global climate as a result of an increase in the amount of greenhouse gases, such as carbon dioxide and methane, in the atmosphere. Changes to the climate may lead to changes in the amount of water stored on the continents, mainly as ice. This will lead to volume changes in the sea, and possibly through thermal expansion owing to altered temperature conditions. During the 20th century, the volume of the sea has increased and the surface of the sea has risen by about 0.1–0.2 cm/year, as established by the world's longest ongoing series of measurements of water level, in Stockholm.

Knowledge of the Sea is Essential

We still have insufficient knowledge of how the oceans, soils, atmosphere and marine and terrestrial ecological systems function both individually and as an integrated global unit (the biogeosphere). We are far from possessing models for the entire biogeosphere, mainly since we lack basic knowledge of many fundamental processes. Under prevailing conditions, there is a large risk that decisions on expensive measures may be reached on the basis of insufficient knowledge, and later found to be without effect and/or unnecessary. On

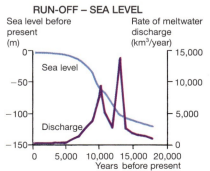

Variations of water level in the sea and melt-water runoff during a glaciation cycle.

Model-calculated concentrations of phosphate and nitrate in the Baltic Proper. Assumptions on how the supply of nutrients will vary in the future enable predictions to be made on how concentrations in the water will change.

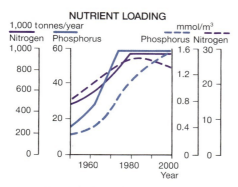

the other hand, necessary measures might not be taken if the problem is not fully understood.

On a regional scale, however, oceanographers have now reached the stage where models are capable of predicting certain effects when, for example, the supply of nutrients into a defined marine area is altered.

Changes in the Future

At present we are unable to quantify possible changes in the future, partly because, as mentioned above, we lack reliable models, but also because we lack knowledge on how conditions will change, e.g., the type and size of the emissions taking place in the future. Nonetheless, we can qualitatively discuss different feasible scenarios. What will the greenhouse effect lead to in the sea? What changes will we experience in the Baltic and the Kattegat because of continued eutrophication and/or continued discharge of toxic substances?

The most probable situation is that, at least on a global scale, there will be no major changes within the foreseeable future. In confined marine areas, however, the situation is different. In areas with a relatively small volume and a limited water exchange with the oceans, such as the Baltic, changes in anthropogenic emissions may almost immediately affect the biochemical status.

CIRCULATION CHANGES

The extremely favourable climate in Sweden in relation to the country's latitude is a result of warm water flowing into the Norwegian Sea over the sill between Greenland-Iceland-Scotland. Low-salinity water from the North Polar Sea flows out along the east coast of Greenland. If wind conditions change, then this water might be dispersed over the entire surface of the Norwegian Sea. The result would be an almost arctic climate in Sweden with long ice-winters and perhaps even with sea ice during the summer. A change from one type of circulation to the other could take place within a short period, within about a decade, and might perhaps take place within the next century.

EFFECTS OF INCREASING WATER LEVEL

The increase in sea level found during the 20th century is considered to be a result of anthropogenic contributions to the greenhouse effect. The increase in water level is expected to accelerate during the next century. On a global scale, the sea level is calculated to increase by perhaps 50 cm up to year 2050, although this figure is very uncertain.

It is mainly in southern Sweden that the effects of the expected rise in sea level will have importance. The isostatic uplift there is the smallest, in fact an isostatic subsidence is reported today from southern Skåne, and large areas of low-lying land are found close to the coast.

A general increase in the water level would allow an increased exchange of water between the Baltic and Kattegat/Skagerrak. This should result in increased salinities in the Baltic which, in turn, would lead to ice formation during the winter being less frequent.

In addition, the greenhouse effect should result in a warmer climate and consequently we may expect reduced problems with icing. The increased salinity will improve the living conditions for cod and other marine animals and plants. The poor oxygen conditions found today in deeper parts of the Baltic will, however, be influenced only marginally. A genuine improvement requires a reduction of the total supply of nutrients from terrestrial and atmospheric sources.

Knowledge of how we influence our environment, both locally and globally, has increased. More people are becoming aware that changes are essential if we are to be able to enjoy the richness offered by our seas and coasts.

Literature och references

Bernes C., 1988. Östersjön och Västerhavet—livsmiljöer i förändring, *Monitor 1988*. Statens naturvårdsverk.

Brasier, M.D., 1980, *Microfossils*, George Allen & Unwin Ltd, London.

Curry-Lindahl, K., 1985, *Våra fiskar—havs- och sötvattenfiskar i norden och övriga Europa*. Norstedt & Söner, Stockholm.

Djurens värld, del 1, 2 och 4, Förlagshuset Norden, Malmö, 1964.

Edler, L., 1980, Planktonalger. *Öresund, tillstånd—effekter av närsalter, Öresundskommissionen*. Liber, pp. 175–204.

Edler, L. and v. Wachenfeldt, T., 1981, *Marina alger, ekologi, dynamik och användning*. Natur och Kultur.

Elmgren, R., 1984, Trophic dynamics in enclosed, brackish Baltic Sea., Réun. Cons.int. Explor. Mer. 183, pp. 152–169.

Eriksson, B., 1982, Data rörande Sveriges temperaturklimat. SMHI Meteorologi och Klimatologi nr 39.

Eriksson, B., 1983, Data rörande Sveriges nederbördsklimat: normalvärden för perioden 1951–80. SMHI Norrköping.

Fairbanks, R.G., 1989, A 17,000-year glacio-eustatic sea level record: influence of glacial melting rates on the Younger Dryas event and deep-ocean circulation. *Nature 342*, pp. 637–642.

Fiskejournalen/Sportfiskaren, *Årsbok nr 12*, 1989.

Fonselius, S.H., 1974, *Oceanografi*. Generalstabens Litografiska Anstalt, Stockholm.

Fonselius, S.H., 1988, Long-term trends of dissolved oxygen, pH and alkalinity in the Baltic deep basins. ICES C.M. 1988/C:23.

Forsman, B., 1982, Evertebrater vid svenska östersjökusten. I *Djur och växter i Östersjön*, Fältbiologerna, Sollentuna.

Franck, H., Matteäus, W. and Sammler, R., 1987, Major inflows of saline water into the Baltic Sea during the present century. *Gerlands Beitr. Geophysics*, Leipzig, vol 96, no. 6, pp. 517–531.

Fritidsfiske–90, 1990. Fiskeristyrelsen, SCB.

Fält, L.-M., 1982, Late Quaternary sea-floor deposits off Swedish west coast. Geologiska institutionen, Göteborgs universitet och Chalmers tekniska högskola, Ser. A, 37.

Förstner, U. and Wittmann, G.T.W., 1983, Metal Pollution in the Aquatic Environment. Springer-Verlag, Berlin, Heidelberg, New York, Tokyo.

Gudelis, V. and Königsson, L.-K. (ed.), 1979, *The Quaternary History of the Baltic*, Acta Universitatis Upsaliensis.

Hainbucher, D., Pohlman, T. and Backhaus, J., 1987, Transport of conservative tracers in the North Sea: First results of a circulation and transport model. *Continental Shelf Res., 7*, pp. 1161–1179.

Hallberg, R.O., 1974, Paleoredox conditions in the eastern Gotland basin during the recent centuries. *Merentutimuslait. Jolk. 238*, pp. 3–16.

Hav–90, 1990, Statens naturvårdsverk.

Jörgensen, B.B. and Revsbech, N.P., 1985, Quantitative estimate of biological mixing rates in abyssal sediments. *J. Geophys. Res., 80*, pp. 3032–3043.

Karlström, C., 1985, Nederbörden i Sverige; medelvärden 1931–60. SMHI Norrköping.

Lindahl O., 1977, Studies on the production of phytoplankton and zooplankton in the Baltic in 1976, and a summary of results from 1973–1976. Medd. från Havsfiskelaboratoriet, Lysekil, no. 220.

Mandahl-Barth, G., 1968, *Vad jag finner på havsstranden*. Almqvist & Wiksell, Stockholm.

Mikulski, Z. (ed.), 1986, Water balance of the Baltic Sea. Baltic Sea Environmental Proceedings, Helsinki Commission, no. 16, pp. 1–176.

Nordberg, K., 1989, Sea-floor deposits, paleoecology and paleohydrography in the Kattegat during the later part of the Holocene. Geologiska institutionen, Göteborgs universitet och Chalmers tekniska högskola, Ser. A, 65.

Nordberg, K., 1989, Giftalger i Kattegatt—en gammal nyhet. *Forskning och Framsteg, 4*, pp. 18–20.

Olaus Magnus Gothus, Rom 1555, *Historia om de nordiska folken*. Gidlund, 1982.

Olausson, E. (ed.), 1982, The Pleistocene/Holocene boundary in south-western Sweden. Sveriges Geologiska Undersökningar, Ser. C, 794.

Pearson, T.H. and Rosenberg, R., 1992, Energy flow through the SE Kattegat: a comparative examination of the eutrophication of a coastal marine ecosystem. Neth. J. Sea Res.

Rosenberg, R., 1982, *Havets liv och miljö*. Liber, Stockholm.

Rosenberg, R. et al., 1990, Marine eutrophication case studies in Sweden. *Ambio 19*, pp. 102–108.

Sjöfartens Miljöeffekter—Inventering och förslag till åtgärder. Sjöfartsverket, 1990.

Stigebrandt, A., 1987, A model for the vertical circulation of the Baltic deep water. *J. Phys. Oceanogr., 17*, pp. 1772–1785.

Thorson, G., 1957, Bottom Communities. Geological Society America, Memoir 67, vol 1.

Thorson, G., 1965, *Livet i havet*. Prisma, Stockholm.

Voipio, A. and Leinonen, M. (ed.), 1984, *Östersjön—vårt hav*. LTs förlag, Stockholm.

Wulff, F. and Stigebrandt, A., 1989, A time-dependent budget model for nutrients in the Baltic Sea. *Global Biogeochemical Cycles, 3*, pp. 53–78.

Wändahl, T. and Bergstrand, E., 1972, Oceanografiska förhållanden i svenska kustvatten. Civildepartementet, Förarbete för fysisk riksplanering, underlagsmaterial nr 18.

Ångström, A., 1974, *Sveriges klimat*. Generalstabens Litografiska Anstalts Förlag, Stockholm.

Authors

Andersson, Lars, 1950, Oceanographer, Swedish Meteorological and Hydrological Institute, Oceanographical Laboratory, Göteborg

Carlberg, Stig, 1945, Senior Oceanographer, Swedish Meteorological and Hydrological Institute, Oceanographical Laboratory, Göteborg

Cato, Ingemar, 1946, Ph. D., Head of Division, Geological Survey of Sweden, Uppsala

Cederwall, Hans, 1944, Ph. D., Dept. of Systems Ecology, University of Stockholm

Edler, Lars, 1945, Ph. D., Swedish Meteorological and Hydrological Institute, Oceanographical Laboratory, Göteborg

Fogelqvist, Elisabet, 1941, Ph. D., University of Göteborg

Fonselius, Stig H., 1921, Ph. D., University of Göteborg

Gadd, Arne, 1920, Author and artist, Göteborg

Grip, Kjell, 1939, Head of Department, Swedish Environmental Protection Agency, Solna

Hernroth, Lars, 1944, Ph. D., Kristineberg Marine biological station, Royal Academy of Sciences, Fiskebäckskil

Håkansson, Bertil, 1950, Senior Oceanographer, Swedish Meteorological and Hydrological Institute, Norrköping

Josefsson, Weine, 1952, Meteorologist, Swedish Meteorological and Hydrological Institute, Norrköping

Josefson, Alf B., 1947, Senior Research Officer, Natural Environmental Research Institute, Roskilde, Danmark

Kjellin, Bernt, 1945, Senior Geologist, Geological Survey of Sweden, Uppsala

Lundqvist, Jan-Erik, 1948, Senior Oceanographer, Swedish Meteorological and Hydrological Institute, Norrköping

Nordberg, Kjell, 1955, Ph. D., Dept. of Oceanography, University of Göteborg

Norrman, John O., 1929, Professor of Physical Geography, University of Uppsala

Omstedt, Anders, 1949, Ph. D., Head of Division, Swedish Meteorological and Hydrological Institute, Norrköping

Rahm, Lars-Arne, 1948, Senior Oceanographer, Swedish Meteorological and Hydrological Institute, Norrköping

Rodhe, Johan, 1943, Ph. D., Dept. of Oceanography, University of Göteborg

Sjöstrand, Bengt, 1939, Ph. D., Institute of Marine Research, Lysekil

Sjöberg, Björn, 1956, Oceanographer, Swedish Meteorological and Hydrological Institute, Oceanographical Laboratory, Göteborg

Stigebrandt, Anders, 1942, Professor of Oceanography, University of Göteborg

Strömberg, Jarl-Ove, 1936, Professor, Head of Kristineberg Marine biological station, Royal Academy of Sciences, Fiskebäckskil

Svensson, Jonny, 1944, Senior Oceanographer, Swedish Meteorological and Hydrological Institute, Oceanographical Laboratory, Göteborg

Tunberg, Björn, 1948, Researcher, Swedish Environmental Protection Agency, Kristineberg Marine biological station, Fiskebäckskil

Vedin, Haldo, 1939, Senior Meteorologist, Swedish Meteorological and Hydrological Institute, Norrköping

Wallentinus, Inger, 1942, Professor of Marine Botany, University of Göteborg, Head of Göteborg Center for Marine Research

Thematic Maps

MAP	SCALE	THEME	PAGE
G1		Bottom sediment in the Atlantic Ocean	9
G2	1:10M	Sea areas	10
G3	1:2,5M	Bottom topography	11–13
G4	1:10M	Bedrock	14
G5	1:2,5M	Sedimentary map	17–19
G6	1:10M	Marine geological maps	18
G7	1:10M	Deposition areas	22
G8	1:10M	Organic chlorine compounds in sediment	24
G9	1:5M	Coastal regions	28
G10	1:10M	Air pressure and wind, January	41
G11	1:10M	Air pressure and wind, July	41
G12	1:10M	Global radiation	43
G13	1:20M	Temperature, January	43
G14	1:20M	Temperature, July	43
G15		Marine forecast areas	44
G16	1:5M	Precipitation	45
G17		Salinity of surface water in the Atlantic Ocean	47
G18		Temperature of surface water in the Atlantic Ocean	47
G19		Surface currents in the Atlantic Ocean	50
G20		Surface currents in the North Sea	50
G21		Tide in the Atlantic Ocean	51
G22		Tide in the North Sea	51
G23	1:10M	Metal contents	54
G24		Distribution of Cesium–137 in surface water, 1986	55
G25		Distribution of Cesium–137 in surface water, 1987	55
G26		Water exchange	56
G27	1:5M	Surface salinity and freshwater supply	57
G28		Drainage area of the Baltic and Kattegat/Skagerrak	57
G29	1:10M	Surface water temperature and ice extent, January 1989	59
G30	1:10M	Surface water temperature and ice extent, April 1989	59
G31	1:10M	Surface water temperature and ice extent, July 1989	59
G32	1:10M	Surface water temperature and ice extent, October 1989	59
G33	1:10M	Maximum extent of sea ice	61
G34	1:10M	Ice thickness	61
G35	1:10M	Ice freeze-up	61
G36	1:10M	Ice break-up	61
G37	1:5M	Wave height	63
G38	1:5M	Water level	65
G39	1:10M	Surface currents and surface water temperature in October	67
G40	1:5M	Oxygen depletion 1990 in Kattegat	68
G41	1:5M	Oxygen depletion 1977 in the Baltic Sea	69
G42	1:5M	Oxygen depletion 1990 in the Baltic Sea	69
G43	1:10M	Inorganic nitrogen and phosphorus	71
G44		Flow of nitrogen	72
G45		Flow of phosphorus	72
G46		Ecosystem, primary and secondary production	77
G47	1:10M	Phytoplankton, number of species	78
G48	1:10M	Phytoplankton, production	78
G49	1:10M	Phytoplankton, amount	80
G50	1:10M	Zooplankton, variation of biomass	81
G51	1:20M	Algae, number of species	84
G52	1:20M	Red algae	85
G53	1:20M	Brown algae	85
G54	1:10M	Macroscopic taxa living on soft bottoms	88
G55	1:10M	Population abundance of macroscopic soft bottom fauna	89
G56	1:10M	Biomass of macroscopic soft bottom fauna	89
G57	1:10M	Phytal fauna, number of taxa	90
G58	1:10M	Phytal fauna, biomass	90
G59	1:10M	Fish species	95
G60	1:5M	Boundaries at sea, 1992	99
G61	1:5M	Middens	100
G62	1:20M	Professional fishermen, 1930	101
G63	1:20M	Professional fishermen, 1960	101
G64	1:20M	Professional fishermen, 1990	101
G65		The development of some Swedish fisheries	102
G66		Distribution of cod	103
G67	1:10M	Cod, Swedish catch, 1989	103
G68		Distribution of herring	104
G69	1:10M	Herring, Swedish catch, 1989	104
G70		Distribution of shrimp	105
G71		Distribution of Norway lobster	106
G72		Distribution of salmon	107
G73	1:10M	Salmon, Swedish catch, 1989	107
G74	1:10M	Fish processing industries, 1935	109
G75	1:10M	Fish processing industries, 1955	109
G76	1:10M	Fish processing industries, 1987	109
G77	1:10M	Fishing occasions and typical catch species	110
G78	1:10M	Rearing rainbow trout, 1989	112
G79		Ferry services, 1989	114
G80	1:10M	Public ports, 1990	115
G81	1:10M	Shipping, 1990	115
G82	1:10M	International trade, 1990	115
G83	1:10M	Proposed sites for wind energy plants	117
G84	1:10M	Reduction of metal emissions	119
G85	1:10M	Iron-manganese concretions	120
G86	1:5M	Mineral resources in the sea	121

Acknowledgements for Illustrations

Permission for distribution of maps approved by the Security Officer. The National Land Survey of Sweden 1992–08–03 and The National Maritime Administration dnr 2905–9250436.

FV = Fiskeriverket (The Swedish National Board of Fisheries)
LMV = Lantmäteriverket (National Land Survey of Sweden)
N = Naturfotografernas bildbyrå (Agency of Nature Photographers/Sweden)
SCB = Statistiska centralbyrån (Statistics Sweden)
SGU = Sveriges geologiska undersökning (Geological Survey of Sweden)
SMHI = Sveriges meteorologiska and hydrologiska institut (Swedish Meteorological and Hydrological Institute)
SNA = Sveriges Nationalatlas (National Atlas of Sweden)
SNV = Statens naturvårdsverk (National Environmental Protection Agency)

Page
2 Photo Hans Ellmén Red algae and starfish at Väderöarna, Bohuslän
6–7 Arne Gadd
8 Photo "Map of the Oceans 'Floor", Tanguy de Rémur. Drawing Hans Sjögren
9 Hans Sjögren
10–11 Map and diagram SNA, data SMHI
12–13 Map SNA, data SMHI
14 Map SNA, data SGU
15 Drawings Nils Forshed, data SGU
17 Map SNA, data SGU/SMHI
18 Map SNA, data SGU
18–19 Map SNA, data SGU/SMHI
20–21 Top Marine geological map Gotska Sandön, SGU 1986, bottom Marine geological map Stora Middelgrund-Halmstad, SGU 1989
22 Map SNA, data SGU
23 Drawings Nils Forshed, from Cato
24 Map SNA, from Cato Photo Torbjörn Lilja/N
25 Drawings Karin Feltzin, from Nordberg Photos Barrie Dale
26–27 Drawings P.O. Nordell, from Norrman
28 Map SNA Photo Tore Hagman/N
29–30 Sverigekartan, SNA Photos LMV
31 Sverigekartan, SNA Photo top LMV Photo bottom John O. Norrman Drawing John O. Norrman
32 Sverigekartan, SNA Photo LMV
33 Sverigekartan, SNA Photos, top and bottom, Arne Philip Photos centre LMV
34 Sverigekartan, SNA Chart 724, 1:50 000, National Maritime Administration 1991 Photo LMV
35 Sverigekartan, SNA Photo top Satellitbild AB Photo bottom LMV
36 Sverigekartan, SNA Photo left Anders Damberg Photo right LMV
37 Sverigekartan, SNA Photos LMV
38 Sverigekartan, SNA Topografiska kartan Umeå 20K NO, 1:50 000 Photo Lars Brydsten
39 Sverigekartan, SNA Photos Göran Albjär Drawing Kerstin Andersson, from Göran Albjär
40 Photo Per-Olav Hoppe Drawing Nils Forshed
41 Maps and diagrammes SNA, data SMHI
42 Diagram Hans Sjögren, data SMHI Photo top Stig Carlberg Photo bottom SMHI
43 Maps SNA, data SMHI Diagram top Hans Sjögren, data SMHI Diagram bottom SNA, data SMHI
44 Photo Per-Olav Hoppe Drawing Nils Forshed, data SMHI
45 Map SNA, data SMHI
46–47 Hans Sjögren
48 Map Hans Sjögren Photo top Lars Andersson Photo bottom Johan Rodhe
49–51 Hans Sjögren
52 Maps Hans Sjögren from Hainbucher, Pohlman and Backhaus Diagram Hans Sjögren from Jörgensen and Revsbech Photo SMHI
53 Photo left SMHI Photo top Stig Westerlund Photo bottom Anders Sjöberg
54 Map SNA, data SNV Drawing Hans Sjögren from Fogelqvist, chart 925, 1:50 000, National Maritime Administration 1991
55 Diagram SNV
56 Map Hans Sjögren, from Monitor 88 and Fonselius Drawing Hans Sjögren
57 Maps SNA, data SMHI Diagram Hans Sjögren, data SMHI
58 Diagram top SNA, data SMHI Diagram centre left Hans Sjögren, from Franck, Matthäus and Sammler Diagram centre right SNA, from Stigebrandt Diagram bottom SNA, data SMHI
59 Maps SNA, data SMHI
60 Diagram SNA, data SMHI Photo top Stig Carlberg Photo bottom SMHI
61 Maps and diagrams SNA, data SMHI
62 Diagram SNA, data SMHI Photo Mats Carlsson Drawing Hans Sjögren
63 Map SNA, data SMHI Photo Björn Larsson/Anders Nolander
64 Photo top Klas Rune/N Photo bottom Claes Grundsten Drawing Hans Sjögren
65 Map SNA, data SMHI
66 Photos SMHI
67–68 Map and diagram SNA, data SMHI
69 Maps SNA, data SMHI Drawing left Hans Sjögren, from Fonselius Drawing right Hans Sjögren, from Hallberg
70 Diagram top Hans Sjögren, from Larsson Diagram bottom SNA, data SMHI
71 Map SNA, data SMHI
72 Maps Hans Sjögren, data SMHI Diagram SNA, data SMHI
73 Photo top Tomas Lundälv Photo bottom left Bo Brännhage/N Photo bottom right Klas Rune/N
74 Photo JGOFS, Woods Hole Oceanographic Institution
75 Photos R.R. Hessler Drawing Liselotte Öhman
76 Drawing top Hans Sjögren Drawing bottom Hans Sjögren/Liselotte Öhman
77 Map Hans Sjögren, from Elmgren, Pearson and Rosenberg
78 Maps SNA, data Lars Edler
79 Photos Lars Edler
80 Map SNA, data Lars Edler Photo Hans Dahlin
81 Map SNA, from Johansson, Kankaala, Lindhal and Hernroth Photo Norman T. Nicoll
82 Photo top Per Jonsson Photo centre and bottom Lars Hernroth
83 Photos Norman T. Nicoll Drawing Vidar Öresland
84 Map SNA
85–87 Liselotte Öhman
88 Map SNA, data SNV Diagram Hans Sjögren Photo Tomas Jangvik
89–90 Maps SNA, data SNV Photo Tomas Jangvik
91–92 Liselotte Öhman
93 Diagram Typoform, from Monitor 88 Drawing Liselotte Öhman, from Thorson
94 Liselotte Öhman
95 Map SNA Drawings Liselotte Öhman
96 Diagram Typoform Photo Bengt Olof Olsson
97 Photo top Klas Rune/N Photo bottom Torbjörn Lilja/N Drawing Hans Sjögren
99 Map SNA, data SMHI/SNV
100 Map SNA, data Jonsson Photo Lars Noord Drawing top Hans Sjögren Drawing centre SNA, from Fredsjö. Drawing bottom Olaus Magnus Gothus, 1555
101 Maps SNA, data SCB Diagram SNA Drawing Hans Sjögren, data SCB
102 Map SNA Drawing Nils Forshed
103 Map SNA, data FV Diagram SNA Photo Bo Brännhage/N Drawing Hans Sjögren
104 Map SNA, data FV Diagram SNA Photo Ingmar Holmåsen/N Drawing Hans Sjögren
105 Diagram top SNA Diagram bottom Hans Sjögren Photo top Ingmar Holmåsen/N Photo centre Bo Brännhage/N Drawing Hans Sjögren
106 Diagram SNA Photo top Bo Brännhage/N Photo bottom Kjell Åshede Drawing Hans Sjögren
107 Map SNA, data FV Diagram SNA Photo Jan Grahn/N Drawing Hans Sjögren
108 Photo Staffan Arvegård/N
109 Maps SNA, data SCB Photo Stieg Eldh Drawing Hans Sjögren
110 Map SNA, data FV/SCB Diagram SNA, data Fiskejournalen/Sportfiskaren Photo Klas Rune/N
111 Photo top Jan Töve J:son/N Photo bottom Alf Linderheim/N Table FV/SCB
112 Map SNA, data SCB
113 Photo top Jonas Sahlin Photo centre and bottom Claes Grundsten/N
114 Photo top Bruno Helgesson/N Photo bottom Göran Hansson/N Drawing Hans Sjögren, data Sjöfartsverket
115 Maps SNA, data SCB and Svensk Lots
116 Photo Bo Sundström
117 Map SNA Photo top Technocean Photo bottom Jan Töve J:son/N
118 Photo top Bo Brännhage/N Photo bottom Tore Hagman/N Table top from Rosenberg, Elmgren, Fleischer, Jonsson and Dahlin Table bottom from HAV–90.
119 Map SNA, data HAV–90 Diagram SNA, data SNV/Boliden mineral Photo Boliden Mineral Table HAV–90
120 Map SNA, from Ingri
121 Map SNA, data SGU and Boström
122 Diagram Hans Sjögren, from Fairbanks Drawing Nils Forshed
123 Diagram Hans Sjögren, from Wulff and Stigebrandt Drawing Nils Forshed

126

Register

abundance 93
abyssal plains 8
air pressure 40 **41 (G10, G11)**
algae 73, 78, 84 **84 (G51)**
algal bloom 79
alum shale 120
amber 121
ammonium 70
anchovies 108
Ancylus Lake 15
anisotropic 53
Appert N. 108
aquaculture 96 112
Archipelago Sea 12
Atlantic Ocean **9 (G1)**, **47 (G17, G18)**, **50 (G19)**, **51 (G21)**

balanid belt 90
Baltic Ice Lake 15
Baltic Proper 10
Baltic shield 14
base line 98
Beaufort scale 41
bedrock **14 (G4)**
Belt Sea 10
benthos 25, 88
biogeosphere 122
biomass 80, 93
bioturbation 93
blue mussels 89
blue-green algae 78, 80
Bothnian Bay 39
Bothnian Sea 12, 36
bottom current 56
bottom fauna 88, 93
bottom sediment **9 (G1)**
bottom topography **11–13 (G3)**
boundary agreements 98
breaking waves 62
brine 48
brown algae 84, **85 (G53)**

cage production 113
cambrian 32
carbon dioxide cycle 76
carbonate mounds 120
catch value 101
Cesium–137 55, **55 (G24, G25)**
chaetognats 83
Chernobyl 24, 55
chlorinated organo-compounds 24, **24 (G8)**
chlorophyll 74, 84
choppy sea 62
ciliates 82
circulation 49
cladocerans 82
cliffed coasts 27
climatic change 122
clouds 42
Coast Guard 98
coastal regions 28, **28 (G9)**
coastal shipping 114
cod 103, **103 (G66, G67)**
concretions 120
condensation 44
continental shelf 9, 16, **17–19 (G5)**, 98
copepods 82, 83
Coriolis force 48, 66
Corophium volutator 89
crust 120
ctenophores 83
currents 66
cyanobacteria 78, **78 (G47)**, 79

Danish Sounds 10
dead bottoms 68, 92
deep water 47
deep water inflow 56
delta 37
delta coasts 27
denitrification 71
density 46, 69

deposit feeders 91
deposition areas **22 (G7)**
deposition bottom 16, 22, 88
diatoms 78, **78 (G47)**, 79
dinoflagellates 25, 78, **78 (G47)**, 79
dispersal patterns 52
drainage area **57 (G28)**

economic zones 98
ecosystem **77 (G46)**
ecosystem model 76
Eem 15
Ekman current 48, 49
Ekman Vagn V. 49
emissions 118
environmental poisons 119
EOCL 24
epifauna 88
epiphyte 84
erosion bottom 22, 88
esker 16
estuary 23
euphotic zone 74
eutrophication 79, 87
excessive fertilization 79, 80

fault 14
fermented Baltic herring 108
ferry traffic 114, **114 (G79)**
fertilizers 118
fetch 62
fish 94
fish meal 104
fish processing industries 108, **109 (G74-G76)**
fish species **95 (G59)**, **110 (G77)**
fish-farming 112
fishing occasions **110 (G77)**
fishing zones 98, **99 (G60)**
fog 42
foraminifers 25
forecast areas **44 (G15)**
freezing-point 48
freshwater supply 56, **57 (G27)**

gales 40
gas prospecting 96
geostrophic current 48, 49
Gislén T. 87
glacial clay 16
glacio-fluvial deposits 16, 39
glaciofluvial sand 33
global radiation 42, **43 (G12)**
Gotland Deep 10
Gradient method 24
gravel deposits 120
green algae 84
greenhouse effect 123
greenhouse gases 76
guano factories 101
Gulf of Bothnia 10
Gulf of Finland 10
gyttja clay 16
Göteborg Center for Marine Research 98

halocline 56
hard bottom 18, 73, 90
haze 42
heavy metals 22
Helsinki Convention 115
herring 104, **104 (G68, G69)**
high water 51
highest shoreline 15, 37
Holocene 15
holoplankton 81
horizontal pressure gradient 48
horsts 32
hydrogen sulphide 68, 74, 75
hypsographic curve 9

ice 58, 60
ice break-up **61 (G36)**
ice cover **59 (G29-G32)**, **61 (G33)**
ice freeze-up **61 (G35)**
ice thickness **61 (G34)**

ICES 102
ichthyoplankton 81
icing 42
increasing water level 123
inertia currents 66, **67 (G39)**
infauna 88
inner waters 98
Institute of Marine Research 98
interglacial 14
internal waves 63
international trade **115 (G82)**
iron-manganese concretions **120 (G85)**
isobar 40, 49
isostatic uplift 27
isotropic 53

Jotnian 14

Karelids 14

Kattegat 10, 30
Kristineberg Marine Biological Station 98
Kylin H. 87

land breeze 40
land elevation 38, 39
land uplift 15
Landsort Deep 10
leisure-time fishing 110, 111
Littorina Sea 15
long waves 63
long-wave radiation 42
low water 51

macroalgae 84
macrofauna 88, 91
macroscopic taxa **88 (G54)**
marinas 115
marine areas **10 (G2)**
marine geological maps **18 (G6)**, 20–21
marine pollution 118
MARPOL 115
marsh ore 120
meiofauna 88, 91
mercury 119
meroplankton 81
metal concentrations 54, **54 (G23)**
metallic discharges **119 (G84)**
metals 118, 119
metamorphosis 83
methane 75
microalgae 86
microfauna 88, 91
microfossil 24
mid-line principle 98
mid-ocean ridge 8
middens 100, **100 (G61)**
mineral resources 120, **121 (G86)**
mixing 53, 56
molecular diffusion 53
monsoon 50
mussels 89, 112, 113

N/P-ratio 71
National Board of Fisheries 98
National Maritime Administration 99
nauplia 83
nitrate 70, 72
nitrite 70
nitrogen 70, 71, **71 (G43)**, **72 (G44)**, 118
nodules 120
North Sea 50, **50 (G20)**, **51 (G22)**
Northern Quark 12, 38
norway lobster 106, **106 (G71)**
Norwegian Channel 50
nutrient limitation 71
nutrient pyramid 76
nutrient supply 118
nutrients 70, 84

oceanic circulation 50
oil 53, 120, 121

oil pollution 119
oil prospecting 96, 116
ores 120
organic material 70
organic substances 53
oxygen 52, 68, 69
oxygen deficiency 68, **68–69 (G40-G42)**, 93
oxygen solubility 68

PAH 24, 53
PCB 53
pelagial 9, 81
peneplane 14, 28, 32
petroleum pollution 24
phosphate 70, **71 (G43)**, **72 (G45)**
phosphorescence 78, 80
phosphorus 70, 71, 118
photosynthesis 68, 74, 84
phytal fauna **90 (G57, G58)**
phytal zone 88
phytoplankton 73, 78, **78 (G47, G48)**, **80 (G49)**
placers 120
plankton 78, 79
polychlorinated substances 53
polycyclic aromatic compounds 24
population regulation 102
ports 114
positioning system 23
postglacial 14
postglacial clay 16
pre-quaternary 14
precipitation 44, **45 (G16)**
primary production 68, 70, **77 (G46)**, 91
production period 74
professional fishermen 101, **101 (G62-G64)**
PSU 46
public ports **115 (G80)**
pulp industry 119

Quaternary 14

radiation balance 42
radioactive elements 55
rag worms 89
rainbow trout **112 (G78)**
recreational boating 115
red algae 84, **85 (G52)**
red tide 80
remineralization 93
residual sediment 18
respiration 68
ripples 16, 26
rotating currents 66
rotifers 82

Saale 15
salinity 46, 58, 88, 95
salmon 107, **107 (G72, G73)**
salt balance 58
sand deposits 120
sand drift 30
sand waves 11
sandy coasts 27
saturated sea 62
scyphozoans 83
sea areas 12
sea boundaries **99 (G60)**
sea breeze 40
sea ice 48
sea smoke 42
sea urchins 90
secondary production **77 (G46)**
sediment 9, **9 (G1)**, 16, **17–19 (G5)**
sedimentation rate 16, 22
seich 64, **65 (G38)**
sewage treatment plants 118
shelf 9
shipping 114, **115 (G81)**
shore fauna 92
shoreline erosion 31
short waves 42, 63

shrimp 105, **105 (G70)**
silica 70
Silurian 14
Skagerrak 10, 29
SMHI 99
snails 89
soft bottoms 73
softbottom fauna **89 (G55, G56)**
Sound 10
Sound bridge 114
sport-fishing 110
sprats 108
stagnation period 56, 68
standing waves 64
starfish 90
Stockholm archipelago 35
Stockholm Center for Marine Research 99
subcambrian 14
succession 78
surface currents **50 (G19, G20)**, **67 (G39)**
surface salinity **47 (G17)**, **57 (G27)**
surface water circulation 50
surface water temperature **47 (G18)**, **59 (G29-G32)**, **67 (G39)**
Svecofennids 14
Swedish Geological Survey 99
Swedish Museum of Natural History 99
swell 63

taxa 88, **88 (G54)**
temperature 43, **43 (G13, G14)**, 47, 60, 66
temperature thermocline 47
terrigenous sediment 9, **9 (G1)**
territorial limit 98, **99 (G60)**
territorial waters 98, **99 (G60)**
Tertiary 14
Theme Water 99
thermocline 58
thunder 42
tide 51, **51 (G21, G22)**, 63
till 16
Tjärnö Marine Biology Station 98
Tornquist Zone 14
trace elements 46, 52, 54
transport bottom 22, 88
trenches 9
tsunamis 63
tunicates 83
turbulent diffusion 53
turnover times 58

Ulvö Deep 12
unstable stratification 44
upwelling 42, 47, 49, 58

vertical mixing 58
volume balance 58

Waern M. 87
water level 51, 64, **65 (G38)**, **67 (G39)**
water masses 46
water turnover **56 (G26)**, 58
wave energy 116, 117
wave height 63, **63 (G37)**
wave length 62
waves 62
Weichsel 15
westerly winds, belts of 50
wind 40, 41, **41 (G10, G11)**
wind energy 96, 116
wind power **117 (G83)**
wind roses 40

Yoldia Sea 15

zonation 84, 88
zooplankton 81, **81 (G50)**

Åland Sea 10

National Atlas of Sweden

A geographical description of the landscape, society and culture of Sweden in 17 volumes

MAPS AND MAPPING

From historic maps of great cultural significance to modern mapping methods using the latest advanced technology. What you didn't already know about maps you can learn here. A unique place-name map (1:700,000) gives a bird's-eye view of Sweden. Editors: **Professor Ulf Sporrong, geographer, Stockholm University, and Hans-Fredrik Wennström, economist, National Land Survey, Gävle.**

THE FORESTS

Sweden has more forestland than almost any other country in Europe. This volume describes how the forests have developed and how forestry works: ecological cycles, climatic influences, its importance for the economy etc. One of many maps shows, on the scale of 1:1.25 million, the distribution of the forests today. Editor: **Professor Nils-Erik Nilsson, forester, National Board of Forestry, Jönköping.** Publication date: **Autumn 1990**

THE POPULATION

Will migration to the towns continue, or shall we see a new "green wave"? This volume highlights most sides of Swedish life: how Swedes live, education, health, family life, private economy etc. Political life, the population pyramid and immigration are given special attention. Editors: **Professor Sture Öberg, geographer, Uppsala University, and Senior Administrative Officer Peter Springfeldt, geographer, Statistics Sweden, Stockholm.**

THE ENVIRONMENT

More and more people are concerning themselves with environmental issues and nature conservancy. This book shows how Sweden is being affected by pollution, and what remedies are being applied. Maps of protected areas, future perspectives and international comparisons. Editors: **Dr Claes Bernes and Claes Grundsten, geographer, National Environmental Protection Agency, Stockholm.**

AGRICULTURE

From horse-drawn plough to the highly-mechanized production of foodstuffs. A volume devoted to the development of Swedish agriculture and its position today. Facts about the parameters of farming, what is cultivated where, the workforce, financial aspects etc. Editors: **Birger Granström, state agronomist, and Åke Clason, managing director of Research Information Centre, Swedish University of Agricultural Sciences, Uppsala.**

The work of producing the National Atlas of Sweden is spread all over the country.

THE INFRASTRUCTURE

Sweden's welfare is dependent on an efficient infrastructure, everything from roads and railways to energy production and public administration. If you are professionally involved, this book will provide you with a coherent survey of Sweden's infrastructure. Other readers will find a broad explanation of how Swedish society is built up and how it functions. Editor: **Dr Reinhold Castensson, geographer, Linköping University.**

SEA AND COAST

The Swedes have a deep-rooted love of the sea and the coast. This volume describes the waters which surround Sweden and how they have changed with the evolution of the Baltic. Facts about types of coastline, oceanography, marine geology and ecology, including comparisons with the oceans of the world. Editor: **Björn Sjöberg, oceanographer, Swedish Meteorological and Hydrological Institute, Göteborg.**

CULTURAL LIFE, RECREATION AND TOURISM

An amateur drama production in Hässleholm or a new play at the Royal Dramatic Theatre in Stockholm? Both fill an important function. This volume describes the wide variety of culture activities available in Sweden (museums, cinemas, libraries etc), sports and the various tourist areas in Sweden. Editor: **Dr Hans Aldskogius, geographer, Uppsala University.**

SWEDEN IN THE WORLD

Sweden is the home of many successful export companies. But Sweden has many other relations with the rest of the world. Cultural and scientific interchange, foreign investment, aid to the Third World, tourism etc. are described in a historical perspective. Editor: **Professor Gunnar Törnqvist, geographer, Lund University.**

WORK AND LEISURE

Describes how Swedes divide their time between work and play, with regional, social and age-group variations. The authors show who does what, the role of income, etc, and make some predictions about the future. Editor: **Dr Kurt V Abrahamsson, geographer, Umeå University.**

CULTURAL HERITAGE AND PRESERVATION

Sweden is rich in prehistoric monuments and historical buildings, which are presented here on maps. What is being done to preserve our cultural heritage? This volume reviews modern cultural heritage policies. Editor: **Dr Klas-Göran Selinge, archeologist, Central Board of National Antiquities, Stockholm.**

GEOLOGY

Maps are used to present Sweden's geology—the bedrock, soils, land forms, ground water. How and where are Sweden's natural geological resources utilised? Editor: **Curt Fredén, state geologist, Geological Survey of Sweden, Uppsala.**

LANDSCAPE AND SETTLEMENTS

How has the Swedish landscape evolved over the centuries? What traces of old landscapes can still be seen? What regional differences are there? This volume also treats the present landscape, settlements, towns and cities, as well as urban and regional planning. Editor: **Professor Staffan Helmfrid, geographer, Stockholm University.**

CLIMATE, LAKES AND RIVERS

What causes the climate to change? Why does Sweden have fewer natural disasters than other countries? This volume deals with the natural cycle of water and with Sweden's many lakes and rivers. Climatic variations are also presented in map form. Editors: **Birgitta Raab, state hydrologist, and Haldo Vedin, state meteorologist, Swedish Meteorological and Hydrological Institute, Norrköping.**

MANUFACTURING, SERVICE AND TRADE

Heavy industry is traditionally located in certain parts of Sweden, while other types of industry are spread all over the country. This volume contains a geographical description of Swedish manufacturing and service industries and foreign trade. Editor: **Dr Claes Göran Alvstam, geographer, Göteborg University.**

GEOGRAPHY OF PLANTS AND ANIMALS

Climatic and geographical variations in Sweden create great geographical differences in plant and animal life. This volume presents the geographical distribution of Sweden's fauna and explains how and why they have changed over the years. There is a special section on game hunting. Editors: **Professor Ingemar Ahlén and Dr Lena Gustafsson, Swedish University of Agricultural Sciences, Uppsala.**

THE GEOGRAPHY OF SWEDEN

A comprehensive picture of the geography of Sweden, containing excerpts from other volumes but also completely new, summarizing articles. The most important maps in the whole series are included. Indispensable for educational purposes. Editors: **The editorial board of the National Atlas of Sweden, Stockholm.**